T0134450

Artificial Intelligence for a Sustainable Industry 4.0

Shashank Awasthi
Carlos M. Travieso-González
Goutam Sanyal • Dinesh Kumar Singh
Editors

Artificial Intelligence for a Sustainable Industry 4.0

Springer

Editors
Shashank Awasthi
G L Bajaj Institute of Technology
and Management
Gr. Noida, India

Carlos M. Travieso-González
University of Las Palmas de Gran Canaria
Las Palmas de Gran Canaria, Las
Palmas, Spain

Goutam Sanyal
National Institute of Technology (NIT)
Durgapur, India

Dinesh Kumar Singh
G L Bajaj Institute of Technology
and Management
Greater Noida, UP, India

ISBN 978-3-030-77072-3   ISBN 978-3-030-77070-9 (eBook)
https://doi.org/10.1007/978-3-030-77070-9

This Springer imprint is published by the registered company Springer Nature Switzerland AG
The registered company address is: Gewerbestrasse 11, 6330 Cham, Switzerland

Dedicated to Covid-19 Warriors

Preface

Artificial Intelligence has become an inextricable part of our society today specifically in the industries where these technologies are being used across sectors. The industries today are undergoing a global renaissance in the field of AI. Sustainability initiatives have thus emerged out as a requirement of this advancement. The implementation of AI poses challenges which can be overcome by adopting certain sustainable methods. This book is hence a good read for having a clear view on the relationship between Artificial Intelligence and Sustainable Industry 4.0.

This book encompasses various topics like Energy Conservation using AI Technology, Industry 4.0, Robotics Process Automation, Communication Technologies, etc. This book provides practical solutions to remain competitive in the age of AI while simultaneously incorporating sustainability, ethical values and future risks. It also dispels common myths about the unethical impact of AI on industries and highlights how these technologies can give way to collaborative leadership.

This book presents valuable researched topics which help in creating a clear understanding as to how the world will efficiently progress by adopting sustainable methods in AI. This book can be a valuable resource for academicians, researchers and professionals working in the field of Artificial Intelligence. It will also contribute to the understanding of the need of a supportive mechanism of sustainable development as the AI technologies improve efficiency and productivity but at the same time pose a danger of imbalance.

Noida, Uttar Pradesh, India   Shashank Awasthi
Las Palmas, Spain   Carlos M. Travieso-González
Durgapur, India   Goutam Sanyal
Noida, Uttar Pradesh, India   Dinesh Kumar Singh

Acknowledgement

Praises to god – the benevolent and the merciful!

To write a well-researched book, proper guidance, direction and support is required, and we are fortunate enough to be able to receive this unconditional support from the erudite fraternity.

We acknowledge the blessings showered by our parents, without whom any task would remain just a dream. We really appreciate the care and patience shown by them at every step.

We acknowledge the effort and contribution of Dr. Upendra Dwivedi and Prof. Anurag Gupta, who have been the strong pillars behind the completion of this book. Our appreciation extends to the entire team that has worked day in and day out in accomplishing the technicalities with ease. We wholeheartedly send my accolades to all of you for your hard work.

Thank You All.

Contents

About the Authors

Shashank Awasthi has a Ph.D. degree in computer science and engineering. He obtained his M.Tech. in computer science and engineering from Dr. APJ Kalam Technical university, Lucknow, and MCA from Dr. BR Ambedkar University Agra. Dr. Awasthi is presently working at G. L. Bajaj Institute of Technology & Management, Gautam Budha Nagar (India), as a professor in CSE Department. His area of interest is WSN and computer networks. He has more than 18 years of teaching and research experience. Dr. Awasthi was invited as plenary speaker to The 4th International Conference on Research in Intelligent and Computing in Engineering 2019 (RICE 2019) held on August 8–9, 2019, at Hanoi University of Industry, Vietnam (HaUI). He has attended and presented his research papers in international conferences. He has published approximately 30 papers in international journals/conference of repute. Dr. Awasthi has membership of both IEEE and the International Association of Engineers, Hong Kong. He has delivered lectures and chaired sessions in internal conferences. He is a member of the editorial board of various reputed international journals. Dr. Awasthi has visited more than eight countries for research purpose. He is certified by Aps Germany, Center for Mechatronics, for industrial robotics under Indo Euro Synchronization.

Carlos M. Travieso-González obtained his M.Sc. degree in 1997 in telecommunication engineering from Polytechnic University of Catalonia (UPC), Spain, and Ph.D. degree in 2002 from the University of Las Palmas de Gran Canaria (ULPGC-Spain). He is Full Professor of Signal Processing and Pattern Recognition and head of Signals and Communications Department at ULPGC, teaching subjects on signal processing and learning theory since 2001. His research lines are biometrics, biomedical signals and images, data mining, classification system, signal and image processing, machine learning, and environmental intelligence. He has researched in 51 international and Spanish research projects, some of them as head researcher. He is co-author of 4 books, co-editor of 24 proceedings books, guest editor for 8 JCR-ISI international journals, and author of nearly 24 book chapters. He has over 440 papers published in international journals and conferences (72 of them indexed on JCR – ISI – Web of Science). Carlos has published 7 patents at the Spanish Patent and Trademark Office. He has been supervisor of 8 Ph.D. theses (12 more are under his supervision) and 130 master's theses. He is founder of the IEEE IWOBI conference series, and president of the steering committee of the InnoEducaTIC conference series and of the APPIS conference series. Carlos is evaluator of project proposals for the European Union (H2020), Medical Research Council (MRC – UK), Spanish Government (ANECA – Spain), Research National Agency (ANR - France), DAAD (Germany), Argentinian Government, and Colombian Institutions. He has been reviewer in different indexed international journals (<70) and conferences (<220) since 2001. He is member of IASTED Technical Committee on Image Processing since 2007 and member of IASTED Technical Committee on Artificial Intelligence and Expert Systems since 2011. He will be ACM-APPIS 2021 General Chair and IEEEIWOBI 2020 and 2020, and was ACM-APPIS 2020 General Chair, IEEE-IWOBI 2019, General Chair APPIS 2019 General Chair, IEEE-IWOBI 2018 General Chair, APPIS 2018 General Chair, InnoEducaTIC 2017 General Chair, IEEE-IWOBI 2017 General Chair, IEEE-IWOBI 2015 General Chair, InnoEducaTIC 2014 General Chair, IEEE-IWOBI 2014 General Chair, IEEE-INES 2013 General Chair, NoLISP 2011 General Chair, JRBP 2012 General Chair, and IEEE-ICCST 2005 Co-Chair. He is associate editor of the *Computational Intelligence and Neuroscience* journal (Hindawi – Q2 JCR-ISI). Carlos was vice dean from 2004 to 2010 of Higher Technical School of Telecommunication Engineers at ULPGC, and vice dean of graduate and post-graduate studies from March 2013 to November 2017. He won "Catedra Telefonica" Awards in Modality of Knowledge Transfer in the editions 2017, 2018, and 2019.

Goutam Sanyal obtained his B.Tech. and M.Tech. degrees from the National Institute of Technology (NIT), Durgapur, and Ph.D. (Engg.) from Jadavpur University, Kolkata, in Robot Manipulator Path Planning. He possesses an experience of more than 35 years in the field of teaching, research, and administration. He has published more than 180 papers in international journals and conferences. Dr. Sanyal has guided 19 Ph.D. scholars of computer science and engineering in the areas of steganography, wireless sensor network, computer vision, and natural language processing. He has guided more than 40 PG and 300 UG theses. He has written a book on computer graphics and multimedia and eight book chapters. Dr. Sanyal is a reviewer of reputed journals (IET, ELSIEVER, SPRINGER, INDER SCIENCE, and Transactions) and conferences. He has filed one patent and has delivered lectures and chaired sessions in internal conferences. He is teaching the following subjects to UG and PG students: computer graphics, computer vision, image processing, VLSI, computer architecture. He visited the University of New Castle and Cranfield in the UK and NUS, Singapore, for research collaboration. He is serving as Ph.D. thesis examiner at VTU, JNTU, Osmania University, Anna University, RPTU Bhopal, and NITs. Dr. Sanyal has served as expert member of faculty selection at NITs and other Universities. He has wide experience in working with different accreditation bodies. He has served as dean of AA & RG, dean of students' welfare, and dean of faculty welfare at NIT Durgapur for nearly 10 years and is at present working as a professor and head of the Department of Computer Science & Engineering. He is a regular Member of IEEE, life member of CSI, and fellow of IEI. Dr. Sanyal has received national awards, and his biography was selected for inclusion in Marquis Who's Who in the World 2016, 2017, and 2018 editions.

Dinesh Kumar Singh received his B.E. in electronics and communication engineering from Kumaon Engineering College, Dwarahat, Almora, in 2003. He has completed his M.Tech. in digital communication at RGPV University, Bhopal, India. He has completed his Ph.D. at the Indian Institute of Technology (ISM), Dhanbad, Jharkhand, India. His area of interest is microwave engineering. Dinesh is currently working as an associate professor in the Electronics and Communication Engineering Department at G L Bajaj Institute of Technology and Management, Greater Noida, UP, India. His research interests include designing of high-gain, compact, re-configurable, fractal-shaped, and circularly polarized microstrip antennas, substrate-integrated wave guide (SIW), and magneto-electric (ME) dipole antenna for modern communication systems. He has published more than 20 papers with various reputed international journals and conferences. He is also reviewer of the *AEU-International Journal of Electronics and Communication* and *Electronics Letter*, among others.

Possibilities of Industrial Trends and Business Benefits with Industry 4.0 Technology: A Survey

M. Arvindhan

1 Introduction

Fortunately, smart manufacturing has now caught on or seen widespread penetration in manufacturing and taken it back to the attention of the general public. In our current age, the factories resemble science fiction, but they inspire, wonder, and are capable of doing many wonderful things.

To expand the domestic manufacturing as well as to keep cost in line with an increasing demand, both of which have been a result of coronavirus disease (COVID), or explicitly cost-cutting and expense repatriation, a significant military stimulus has arisen for obvious reasons. A connected world of the Internet of Things is shown in Fig. 1. As company assets, computers, networks varying from wireless low-powered networks to connected high-capacity networks connected with one another, the number of system-underlying sensors, and subsystems are brought online, the volume of data processed grows, and it allows us to use mobile solutions like smartphone, notebooks, wearables, and a wearable tech device or a smart headband [1, 2].

Virtualization and cloud computing include the use of lower cost processing and storage, too. Used in many industries, particularly in energy based and the cyber-physical systems (CPSs) for data processing is monitoring and managing physical processes and IoT physical infrastructures, with sensor, actuator, and chip-based digital modeling. These are also the same type of CPSs. These big data analytics and IoT workflows are invaluable in predicting, supporting, creating, and monitoring innovative manufacturing, also known as robots and 3D printers. A good question, indeed, but it's a question that causes the industrial specialist to falter in their explanations; e.g., those entangled in information get bogged down by this concept. You

M. Arvindhan (✉)
Galgotias University, Greater Noida, UP, India
e-mail: m.arvindhan@galgotiasuniversity.edu.in

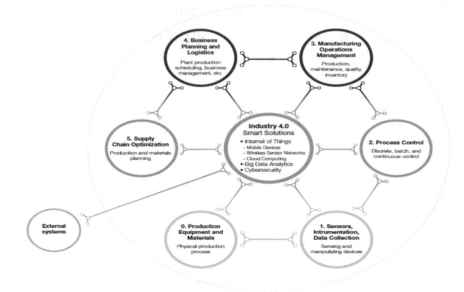

Fig. 1 Smart solution for Industry 4.0

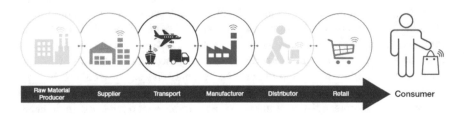

Fig. 2 Positive effect of smart production benefit to our industry

ought to be extra cautious about naming the list, as the number of "benefits" is a dozen, and any order you give them does not reflect how it really appears to a benefit-lay people. Other people view them as a true tech person's guide to the new and greatest cool technology, but it sounds more like a project to be thrown in a pool of research to most [3, 4].

The best concise, simplistic, obvious approach can be answered by grasping the audience's point of view; from where they are posing the question, the issue comes from. They inquire about the consumers' side or the general public's side (Fig. 2).

2 How the Concept of Supply and Demand Applies to the Overall Economy?

The most valuable advantage is that you will be able to order a single piece to your individualized needs without incurring any extra costs because mechanical machinery is manufacturing them [5, 6].

To reiterate, this is a direct advantage, not a capability or a technology. While mass-produced and planned things have become the norm in our culture, the simplicity of a product designed for us and made especially for our use is quick to grasp.

In the long term, those who have the potential to deliver goods that consumers want on a made-to-order basis will usurp those who cannot keep up with new technical advances and will get a large market share. Amazon's strong hold on the retail industry has essentially been built by a deliberate lack of effort to catch up to those who have moved slower. In the United States, things are beginning to get close on the manufacturing side as well.

Second, the ability to create fast, flexible, and nimble production facilities means that you have a lot of data and information readily available. In addition to the data availability, those data serve as the basis for personalizing the commodity; it lays the foundations for a predictive maintenance system [7, 8].

For several decades, predictive maintenance has become a mostly fictitious goal; however, even though it has been widely touted, it has so far been absent in business. The big challenge with predictive analytics is the impact it has on minimizing downtime (or failure). Not only it is incredibly costly, but downtime also increases system repair costs.

And downtime is supposed to be avoided by operational predictive maintenance measures, as well as those that have been developed over the years to ensure it doesn't occur. As beneficial as it is, preventative maintenance often has a high associated cost. Instead of reacting quickly to specific problems with maintenance actions, Industry 4.0 adopts machine-to-machine (M2M) connectivity and intelligent equipment that enables proactive analytics to become a reality [9, 10].

The savings accrued in a scheme falls to the bottom line, but there's no question they'll be made in the offices where technical people (those who know less about the money side of things) use them quickly, which is a direct advantage (and not merely a geeky feature).

To me, those are the Big Three: What smart production is all about Industry 4.0 empowers us to consider the significance of Industry 4.0 on both the customer and business sides—that a business investment would be smart. Industry 4.0 has been outlined in recent publications by Cooper and James [11] and Dominici et al. [12] using a process called a literature review (Fig. 3). Four critical components were singled out in the review: the Internet of Things, the Internet of Services (IoS), the cyber-physical systems, and smart factory. Formal relationship: Smart products and machine-to-machine networking are not separate components in Industry 4.0.

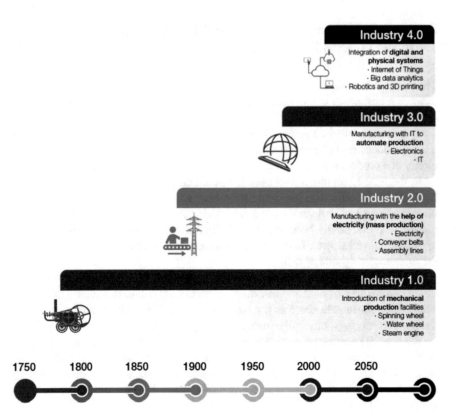

Fig. 3 Revolution on the industrial sector in periodic

The direct initiators of CPSs are the innovative products and Internet of Things. Process is often referred to as "smart products," "Inn-Industry," or "integrated industry."

3 Technology Road Map for Industry 4.0

New manufacturing methods using Industry 4.0 with the Internet of Things, cyber-physical networks, cloud computing, and big data have the ability to get stronger and generate new goods and ideas. It self-optimizes.

The below road map demonstrates the growth direction of Industry 4.0. There are four distinct phases. The songs, music, etc. they have written and produced; the albums, music, etc. they have made digital newcomers: They are on the road to introducing Industry 4.0. Original digitization outcomes are met in sections as well as discrete commodity portfolios. Meeting requirements is not assured in this point, and electronic hazards are not tracked. The second level of Industry 4.0 businesses is a vertical integrator. Many corporations still use advanced devices (also known as

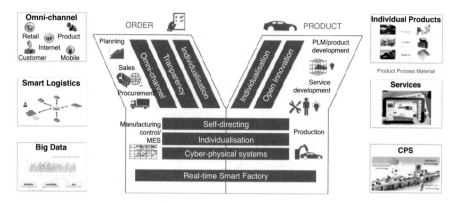

Fig. 4 Industry 4.0: the big change

embedded systems), which enables the delivery of products over the Internet. The presence of the network offers consumer arrival through overt web sites and inventory (Fig. 4).

3.1 Smart Factory

They are all so cyber-physical systems (CPSs) using I4.0 technologies. These software-intensive devices are linked to the Internet and are capable of communicating with each other as well with non-software-based materials. These materials are said to be intelligent if they have characteristics such as consistency and production processes within them (chip). Using read-frequency identification (RFID) technology, these materials can pass through the production process without relying on man-made points of reference. If a pump goes awry, or if another system takes over, a CPS platform or set of materials can execute coordination across a quasi-marketplace. A noteworthy feature is the storage efficiency of massive amounts of data, as a result of reducing the price of storage media and improvements in "in-memory" database technologies (how important storing of data in computer memory, thanks to low storage costs and quicker computer memory technology) [13–15].

In an I4.0 setting, all of the products' life cycle operations such as repairs, servicing, modifications, and application requirements are automated.

3.2 Cargo Loading and Unloading

This ensures that the user interface has to be dynamically adjusted to the environment. Individualization and improved access to a supplier bring greater instances of transition and thus increase versatility in design and manufacturing. The buyer has the ability to modify the initial specification, for example, the color of his car up to the very beginning of the production period [16, 17].

Platform orientation means that in any case a web service is made available. This reflects a deliberate change from vast enterprise resource planning (ERP) structures to monolithic structures. Smaller device units exist, such as software applications, which can be manually modified and reconfigured by the user.

3.2.1 Formal integration platform

Integrating dissimilar device platforms is feasible through the use of a model-based system. A workflow platform offers usable facilities. Both systems consist of various platform modules that are put together into larger applications (Fig. 5).

The fields on which a company wishes to concentrate can differ, based on any of the three possible purposes: whether it is for resolving problems, to satisfy consumer needs, or to boost productivity and continuity. This may vary from a manufacturing cell or a tool, each of which could be used to measure power, output, production, or energy consumption. Detailed information on this will be given in more depth in a little while [18–20]:

- Regardless of the region, we suggest following a three-step method.
- Integration into Industry 4.0.
- The use of sensors and switches.
- Improving the sensors' practical capability.
- 100% rollout, in which Industry 4.0 capabilities are provided in the farm.
- Sensors and controls in output.

The IoT gateway includes sensors, which are built on Linux operating systems and open interfaces as well as many Java applications. Even, before you mount this into a farm, decide the specifications or goals to be accomplished with the IoT gateway. In the long run, this will be the sensors' true potential location: the commercial IoT portal—which is important to the Internet of Things, providing a cost-effective and flexible way for companies to connect and link together new and current hardware in order to streamline their processes and enhance product quality.

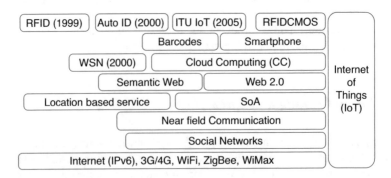

Fig. 5 Connection between IoT with Industry 4.0

3.2.2 Technological Challenges and Accompanying Ranging of Data

The four features of Industry 4.0 further show the significant potential for improvement in both industrial and conventional industries: vertical systems networking, modern corporate supply chain, horizontal convergence in the value chain, and exponential developments.

3.2.3 Vertical Incorporation of Intelligent Development

Industry 4.0's most important feature is the vertical networking of smart manufacturing networks in the factories. This vertical networking leverages cyber-physical systems (CPSs) to allow plants to respond quickly to fluctuations in demand or fault condition. Customer-tailored and individualized output integration is needed for this. Proprioceptive sensor technology also helps in detection and autonomous management. CPSs enable total self-sufficiency in manufacturing as well as administration. Resources and goods can be interconnected, and inventory can be stored anywhere. Every step in the process is recorded, and errors are detected automatically. Amendments to orders should be dealt with promptly; variations in quality or equipment breakdowns can be managed easily. Such processes also enable wear and tear on the materials to be monitored more effectively or preempted. Overall, excess has been cut [21].

3.3 A New Wave of Corporate Supply Chains

Another important feature of Industry 4.0 is horizontal convergence of multinational value-chain networks. These value-creation networks make use of real-time performance and allow both rapid problem-solving and global optimization.

Similar to shared distribution networks, they include networking all the way to warehousing, promotion, and advertising. All of the past is tracked and retrievable at any time, thus promoting "product memory."

This helps expand the openness and interconnectedness of whole process chains, for example, from buying to sales and back. Dynamic product-specific solutions may be rendered not only throughout the manufacturing, preparing, composing, and distributing phases, but also with regard to consistency, time, price, and environmental security in real time along the value chain.

Integration through consumers and suppliers threatens to create entirely new business models, and it is capable of generating entirely new business models for those involved. Judicial concerns and problems about liabilities and exclusive rights are growing in significance [22].

3.4 All the Way Through the Supply Chain

The third core attribute of Industry 4.0 is the integrated use of cross-disciplinary knowledge and operations across the whole value chain and with consumers from initial concept creation to end-of-life of all products and services.

The engineering involved in the design, production, as well as the manufacturing of new products and services is flawless. New or updated systems are needed for the introduction of new products. The creation and manufacturing of new goods are also tied to life cycles, meaning new synergy can be formed between them [23].

Representative of this is that data and knowledge exist during the commodity life cycle, allowing modern, more agile systems to be identified as data, as well as at all times.

The technological advancement and increase in productivity are possible due to the use of exponential technology. Industry 4.0 incorporates the effects of exponential technology and personalized methods. Currently, Industry 4.0 platforms need cognitive and highly autonomous automated solutions (Fig. 6). Artificial intelligence (AI), robots, and highly developed sensor technologies have the ability to go much further toward being highly autonomous and instant-flexible.

Supply chain management (SCM) can save time and money and can help to prepare driverless car routes in factories and warehouses. Effective nanomaterials and nanosensors could also be used in quality monitoring and the manufacturing of next-generation robotics for healthy human–robot interaction. A method of flying repair robots in manufacturing facilities at any time of the day and in any atmosphere is the one which would be only possible in autonomous factories [24].

The flexibility that 3D printing is demonstrating here is a perfect example of how I4.0 exponential technologies are moving industries (e.g., additive manufacturing), operational models (e.g., modern design/build capabilities), and enhanced supply chain operations (e.g., disintermediation of supply chain members, customer integration). Overall, more attention would be paid to maintaining quality control or modifications and on-location printing of spare parts requirements. The rules of

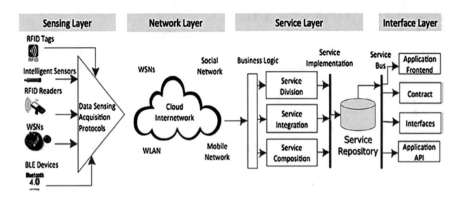

Fig. 6 The four characteristics of Industry 4.0

intellectual property, product liability, customs duty, and value-added tax have yet to be determined. Although 3D printing now exists for all materials (including metals, plastics, ceramics, and living cells), few are suited to industrial needs with regard to pore structure and other properties. Where the requisite standard of consistency has already been reached, protracted content certification procedures are in place. Cyberattacks and viruses have the ability to put smart processing processes to a halt, which may cost significantly. For the most part, in our interviews, it was manufacturing executives who emphasized that risks would probably be handled by a personalized risk control and an acceptable security plan. Decentralized systems, modular safeguards, and few privileges were found to be a good compromise in this case. They also expected that improvements in the essence of information defense would not outpace the growth of the threats. Many respondents also emphasized the need of modern and binding data protection policies and practices [25, 26].

4 Aspects for IT Management to Consider

The underlying industrial utility of 3D printing is the sense of the industry. Additive processing also goes by the name of 3D printing. The technologies deal with the production of solid blocks rather than the removal or milling of solid materials.

Lithographic manufacturing takes place by the sequential addition of powder or oil. The substances include various metals, fibers, composites, and polymers. There are four process categories, all using additive processing or technology [27].

Stereo lithography, digital light processing (DLP), or digital film imaging can harden a polymer with light to extrusion deposition; the melted plastic is applied in layers using a hot-glue gun. Granular grinding of a powder is performed with a printer head or laser. Laser additive processes include laser sintering, selective laser melting, metal laser sintering, metal and electron beam melting, and additive 3D printing. A part is added one layer at a time in a laminated object manufacturing (LOM) phase. New developments still have a high investment cost of production and implementation and deployment, and the 3D printing technology is no different. At the outset, the initial cost of each manufactured unit is high, but when more units are made, the cost goes down. It is a significant initial investment to set up a factory, but in mass production, it is possible to recoup the costs over time. Using additive technology isn't really inexpensive at the beginning; you have to spend a significant sum of money to start out-of-pocket (although the cost of tools is lower). In comparison with traditional manufacturing, the prototypical cost curve also flattens [28].

5   Concepts of Creativity

Integrated and cross-disciplinary innovation will emerge during the value and product life cycles in Industry 4.0. The purpose of Industry 4.0 applications is to promote product innovation only.

The main opportunity for innovation for most companies lies in three aspects: product offerings, operational systems, and network-based processes, as well as customer-facing roles, such as branding and customer engagement (as described in the Deloitte, monitor 10 forms of innovation) [29].

Informal analysis indicates that the share price of firms employing only two strategies of invention does well on the capital market. 3D printing is an important feature for Swiss manufacturers. Half of the respondents believe 3D printing is critical to Industry 4.0.

Integrated and cross-disciplinary innovation will emerge during the value and product life cycles in Industry 4.0. The purpose of Industry 4.0 applications is to promote product innovation only.

The main opportunity for innovation for most companies lies in three aspects: product offerings, operational systems, and network-based processes, as well as customer-facing roles, such as branding and customer engagement (as described in the Deloitte, monitor 10 forms of innovation) [30, 31].

Informal analysis indicates that the share price of firms employing only two strategies of invention does well on the capital market (Fig. 7).

6   The Effective Administration of Creativity

A sound approach to innovation strategy can be executed across the board, including initiative, business, and product management. By virtue of the digital revolution, it would be possible to boost Industry 4.0 productivity in all these regions. Platform and individualized curricula promote customized preparation, enabling adoption and organization growth to occur more rapidly. In Industry 4.0 projects, tracking both the return on investment (ROI) and the risks becomes simpler. Information technologies (ITs) can be used to help accelerate the pace of research and development (R&D) and transform the knowledge exchange of global networks under the same principles that online gaming culture platforms employ [32].

The transition to Industry 4.0 would allow relevant data to be available at any time and in any place. These data will have both the observations and effects of big data processing artificial intelligence (AI) (Fig. 8). Global cross-checking and base-basing of decisions based on applicable data. It will aid businesses in understanding and meeting their consumers' desires, while also fine-tuning the production cycles.

Life implementations and for the evolution of virtual reality (VR) in modern life uses. Improving human output is the objective of augmented reality (AR), providing the necessary knowledge for a specific mission. This novel tool offers numerous

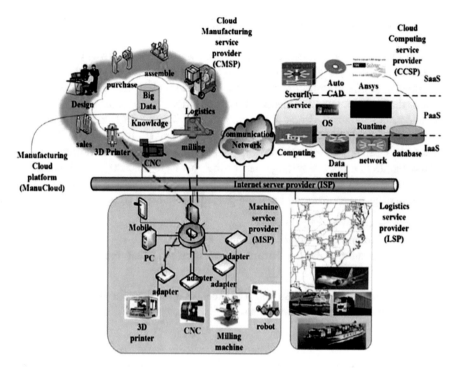

Fig. 7 Industry 4.0 promotional products

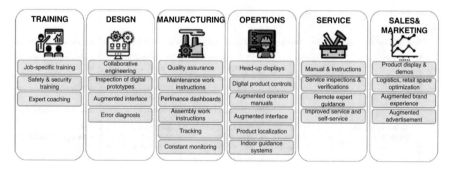

Fig. 8 Life cycle management in production using Industry 4.0

effective solutions, aiding the work as an human machine interface (HMI). These technologies are used in many different industries, for example, amusement, marketing, recreation, repair, engineering, logistics, and so on. Currently, AR technology is used in a wider range of industrial settings. The use of AR in manufacturing processes for simulation, assistance, and advice has been shown to be effective. Formal science involves the use of artificial intelligence about the natural world, where knowledge itself is what's in use. The usefulness of AR is limited only by how it deals with human senses. The use of AR will fill in holes such as between product production and manufacturing [33].

The theory of AR can be described as follows: Two alternatives converge into one. The method of digitally manipulating reality to create two-dimensional (2D) flat objects is often referred to as the use of 2D digital object processors or other claims that only look at 2D objects. Authors described these features as (1) the ability to combine real and virtual objects in a real environment, (2) the ability to communicate with and to align virtual and 3D objects, and (3) the ability to operate in real time. Utilizing traditional hardware has the use of zero or very nominal outlay. Some people find the see-through part of the glasses to be costly. The type of networking provided to the area, e.g., military versus commercial, could be chosen based on differences, to avoid breaking procedures [34, 35] (Fig.9).

1. Wearable and in everyday use.
2. Spatial, geometrical (projector and hologram).
3. Processing unit: delivering virtual information to arouse the show of interactive facts.

Examples include quick response (QR) sensors, global positioning system (GPS) locators, digital photographs, and sensors.

Leadership	Create awareness and discover the\ benefits	Build proof of concepts Collaborats with IT and OT vendors	Define/implement an Industry 4.0 strategy for the enterprise	Define/implement an Industry 4.0 strategy for the value chain	Create new (disruptive) products, services and business models
Data	Data silos	Share data between departments analyze, and obtain new insights	Share data across the enterprise Implement Master Data Management	Share data between companies implement clear data ownership	Share data across the value chain implement data quality as KPI
Technology	Technology silos Descriptive analysis	Connected devices and data sources Diagnostic analytics	IT modernization Application integration IoT platform	Apply secure open standards Predictive analytics	Xaas Prescriptive analytics
Security	Basic network protection	Ad hoc application and infrastructure protection	Enterprise-wide application and infrastructure protection	Identity-aware information protection	Adaptive and automated security control
	Ignoring	**Exploring**	**Standardizing**	**Adopting**	**Adapting**

Fig. 9 Industry 4.0 characteristics based on maturity level

The process of managing the flow of sensitive information and data to ensure the confidentiality, integrity, and availability of an online information system.

Security has to be maintained across the supply chain and integrated across manufacturing locations. I4.0 technology must provide cyber security to learn from it. A direct assault on industrial control systems (ICSs) by malicious people or programs can be perilous. These industries are controlling processes such as Supervisory Control and Data Acquisition (SCADA) and hierarchical management systems [36].

1. It is concerning to note that devices have not been using protection or anti-virus software for too long (weeks/months) without patches or updates.
2. A significant number of ICS controllers are still in use in an ICS prior to computer system (CS) being a problem.
3. Several CS networks will join the CS bypasses since there are various routes from the different ICS networks.

Computing is the union of "electric and computer devices of cyber." The "digital component" creates a direct interaction between the "physical component" (such as mechanical systems) and the "real world" by making a simulated duplicate of it. This computer representation would include the "real world" knowledge and data (i.e., a cyberization of the CPS). As a consequence, computational power management (CPM) may be believed to be a collection of transformative technology (Fig. 10). The CPSs utilize augmenting their intelligence and ability to undertake unique or nonrepetitive ventures, but may not perceive themselves as accepting smart components. For instance, this piece of equipment controls the demands of the workpieces, adjusts manufacturing techniques for maximum performance, and makes (or finds) new ones if appropriate [37]. Their network is now in operation. As a consequence, the CPS is the physical environment's embedded device. This machine would carry out activities that were carried out by dedicated computers. The CPS model can be thought of as a control device that contains one or more microcontrollers, which communicate with the environment by controlling sensors and actuators as well as process data acquisition and processing (Fig. 11). This

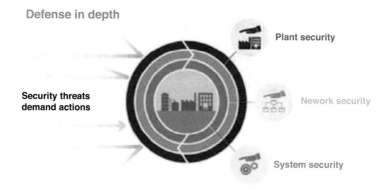

Fig. 10 Security model designing in Industry 4.0

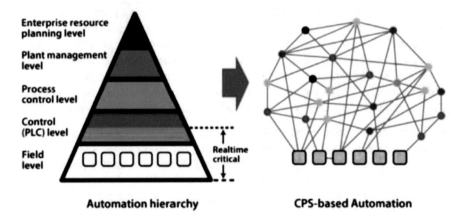

Enterprise resource
planning level

Plant management
level

Process
control level

Control
(PLC) level

Field
level

Realtime
critical

Automation hierarchy　　　**CPS-based Automation**

Fig. 11 Hierarchy decomposition of the traditional automation pyramid and the CPS approach

device will be equipped with a communication interface to allow it to share data with the cloud or with other systems. The term "CPS" is widely used in the sense of the IoT definition. According to Humayun et al., the CPS comprises three elements: (1) communication; (2) computing and control; and monitoring; and (3) avoidance.

7 Encompasses Every Facet of Our Lives from Social Media to E-Commerce

Replacing physical objects for the Internet of Services (IoS) uses the premise that new value-added services are accessible over the Internet and allows individuals and organizations to create, package, and deliver their own Internet services. Service providers may use the Internet of Things as a means to deliver their services. Thus, the commodity-oriented production industry is increasingly transitioning to service-oriented to serve during the product life cycle. In this respect, consistency means the same as "the highest possible."

Products are allowed by service-oriented architecture (SOA), which also has a strong competitive advantage by value-added services. The consistency of the experience is enhanced because IoS collects product details, e.g., during activity or for new service production. Business divisions and teams form their own strategic and job schedules. Each machine and each database have their own silo running in parallel. The goal is to streamline the job performed by the operational team.

7.1   Collecting Data

It is used for investigating and addressing "what occurred?" Production facilities and devices are not connected to the Internet for network security purposes. Individuals must be conscious, interested, and knowledgeable in order to advance to the next stage of maturity [38].

7.2   Looking out Opportunities

Company units and agencies shape alliances. Encouraging them to work together would open up new areas of discovery. In data analytics, data processing and interpretation were introduced to address the query "why did it happen?" Data from the operational technology (OT) and IT IoT systems are being collected and analyzed. Industry 4.0 Evangelists launch the first programs. Application and networks are made resistant to threats using ad hoc authentication. Mechanisms are given to clients and servers [39].

7.3   Consistency

Business units and roles work together in the organization to standardize. Additionally, the company describes a business data and method infrastructure as well as a policy and reference architecture. Equipment and services are connected to the Internet of Things prediction, resolving the query "what will happen in the future?" Enterprise and infrastructure are protected.

Collaboration between market environments and the company then describes the Industry 4.0 business environment approach and reference architecture. Data are spread across business borders. In order to ensure real-time data collection and analytics are put to use in activities, operators need to monitor "what is occurring?" Applications based on the Internet of Things systems have extra features and statistical models to aid decision-making [40]. The five businesses in this environment are well developed and lead the industry landscape. They launch new and innovative business models, which include an automated supply chain. Real-time data analysis and data analytics are used in activities to decide how to make things possible. Formal models are used to predict and to simplify company operations. There are defense mechanisms that are both flexible and automatic.

8   Conclusion

The entire industrial environment will be disrupted with Industry 4.0, from design and corporate structure to products, facilities, and even market models. These ideas may be developed and implemented incrementally, but it is only a matter of time before they become a larger part of mainstream. Companies who ignore emerging technology and invest in pilot projects will be at a severe disadvantage and be left behind in the industrial revolution currently occurring in the sector. Industry 4.0 is the technological transformation of the twenty-first century, where decreases in prices and increased productivity are contributing to "smarter" goods and services and human capitals are a requirement for usage. The systems' automated operations, simplistic setup, basic needs, and overall increased reliability that boost productivity, as well as the high scalability of the solution, have the potential to help businesses in the manufacturing industry to simplify their processes and drive overall business results.

References

1. D. Kolberg, D. Zuhlke, Lean automation enabled by industry 4.0 technologies. IFAC PapersOnLine **48**(3), 1870–1875 (2015)
2. F. Zezulka, P. Marcon, I. Vesely, O. Sajdl, Industry 4.0 – An introduction in the phenomenon. IFAC PapersOnLine **49**(25), 8–12 (2016)
3. J. Qin, Y. Liu, R. Grosvenor, A categorical framework of manufacturing for Industry 4.0 and beyond. Procedia. CIRP **52**, 173–178 (2016)
4. C.J. Bartodziej, *The Concept Industry 4.0 – An Empirical Analysis of Technologies and Applications in Production Logistics* (Springer Fachmedien Wiesbaden, Wiesbaden, 2017)
5. Ustundag A, Cevikcan E (2018) Industry 4.0: Managing the Digital Transformation. Springer series in advanced manufacturing
6. A. Rojko, Industry 4.0 concept: Background and overview. Int. J. Interact. Mb. Technol. **11**(5) (2017)
7. D. Gorecky, M. Schmitt, M. Loskyll, D. Zuhlke, Human- machine-interaction in the industry 4.0 era, in *2014 12th IEEE International Conference on Industrial Informatics (INDIN)*, (2014), p. 14718826
8. A. Varghese, D. Tandur, Wireless requirements and challenges in industry 4.0, in *Proceedings of 2014 International Conference on Contemporary Computing and Informatics, IC3I 2014*, (2015), pp. 634–638
9. F. Shrouf, J. Ordieres, G. Miragliotta, Smart factories in industry 4.0: A review of the concept and of energy management approached in production based on the internet of things paradigm publisher: IEEE, in *2014 IEEE International Conference on Industrial Engineering and Engineering Management*, (2014), p. 14983686
10. A. Angelopoulos, E.T. Michailidis, N. Nomikos, P. Trakadas, A. Hatziefremidis, S. Voliotis, T. Zahariadis, Tackling faults in the industry 4.0 era—A survey of machine-learning solutions and key aspects. Sensors **20**(1), 109 (2020)
11. J. Cooper, A. James, Challenges for database management in the internet of things. IETE Tech. Rev. **26**, 320–329 (2009)

12. G. Dominici, V. Roblek, T. Abbate, M. Tani, "Click and drive": Consumer attitude to product development. Towards future transformations for driving experience. Bus. Process. Manag. J. **22**, 420–434 (2015)
13. M. Marolt, Z.D.H. PuciharA, Social CRM adoption and its impact on performance outcomes: A literature review. Organ **48**, 260–271 (2015)
14. V. Roblek, M. Bach, M. Mesko, A. Bertoncelj, The impact of social media to value added. Kybernetes **42**, 554–568 (2013)
15. M. Rodriguez, K. Trainor, A conceptual model of the drivers and outcomes of mobile CRM application adoption. J. Res. Interact. Mark. **10**, 67–84 (2016)
16. H. Kagermann, Change through digitization – Value creation in the age of industry 4.0, in *Management of Permanent Change*, (Springer, Wiesbaden, 2015), pp. 23–45
17. J. Yu, N. Subramanian, K. Ning, D. Edwards, Product delivery service provider selection and customer satisfaction in the era of Internet of Things: A Chinese e-retailers' perspective. Int. J. Prod. Econ. **159**, 104–116 (2015)
18. Hessman, T. The down of a smart factory. Technical report, Industry Week. https://www.industryweek.com/technology-andiiot/article/21959512/the-dawn-of-the-smart-factory   (2013). Accessed Feb 2020
19. European-Commission. Program smart cities. Technical report, The EU. https://ec.europa.eu/energy/en/content/programsmart-cities (2015). Accessed Feb 2020
20. Proctor, M., Wilkins, J. 4.0 sight—Digital industry around the world. Technical report, EU Automation. http://www.4sightbook.com/ (2019). Accessed Feb 2020
21. P. De Silva, P. De Silva, Ipanera: An industry 4.0 based architecture for distributed soil-less food production systems, in *Proceedings of the 1st Manufacturing and Industrial Engineering Symposium, Colombo, Sri Lanka*, (2016)
22. K. Zhou, T. Liu, L. Liang, From cyber-physical systems to industry 40: Make future manufacturing become possible. Int. J. Manuf. Res. **11**(2), 167–188 (2016)
23. M. Mladineo, I. Veza, N. Gjeldum, Solving partner selection problem in cyber-physical production networks using the HUMANT algorithm. Int. J. Prod. Res. **55**(9), 2506–2521 (2017)
24. R. Baheti, H. Gill, Cyber-physical systems, in *The Impact of Control Technology: Overview, Success Stories, and Research Challenges*, (IEEE Control System Society, New York, 2011), pp. 161–166
25. E.A. Lee, Cyber physical systems: Design challenges, in *Proceedings of the 11th IEEE Symposium on Object/Component/Service-Oriented Real-Time Distributed Computing*, (Piscataway, The Institute of Electrical and Electronics Engineers, 2008), pp. 363–369
26. T.H.J. Uhlemann, C. Lehmann, R. Steinhilper, The digital twin: Realizing the cyber-physical production system for industry 4.0. Procedia. CIRP **61**, 335–340 (2017)
27. A.C. Valdeza, P. Braunera, A.K. Schaara, A. Holzingerb, M. Zieflea, Reducing complexity with simplicity-usability methods for industry 4.0, in *Proceedings 19th Triennial Congress of the IEA*, vol. 9, (2015, August), p. 14
28. Vijaykumar, S., Saravanakumar, S. G., & Balamurugan, M. (2016). Unique sense: Smart computing prototype for industry 4.0 revolution with IOT and bigdata implementation model. arXiv preprint arXiv:1612.09325
29. J. Wan, S. Tang, Z. Shu, D. Li, S. Wang, M. Imran, A.V. Vasilakos, Software-defined industrial internet of things in the context of industry 4.0. IEEE Sensors J. **16**(20), 7373–7380 (2016)
30. K. Wang, Intelligent predictive maintenance (IPdM) system– Industry 4.0 scenario. WIT Trans. Eng. Sci. **113**(1), 259–268 (2016)
31. S. Wang, J. Wan, D. Zhang, D. Li, C. Zhang, Towards smart factory for industry 4.0: A self-organized multi-agent system with big data based feedback and coordination. Comput. Netw. **101**, 158–168 (2016)
32. A.A.F. Saldivar, Y. Li, W.N. Chen, Z.H. Zhan, J. Zhang, L.Y. Chen, Industry 4.0 with cyber-physical integration: A design and manufacture perspective, in *21st International Conference on Automation and Computing (ICAC), 2015*, (IEEE, 2015, September), pp. 1–6

33. R. Schmidt, M. Möhring, R.C. Härting, C. Reichstein, P. Neumaier, P. Jozinović, Industry 4.0-potentials for creating smart products: Empirical research results, in *International Conference on Business Information Systems*, (Springer, 2015, June), pp. 16–27
34. G. Schuh, T. Gartzen, T. Rodenhauser, A. Marks, Promoting work-based learning through industry 4.0. Procedia. CIRP **32**, 82–87 (2015)
35. G. Schuh, T. Potente, C. Wesch-Potente, A.R. Weber, J.P. Prote, Collaboration mechanisms to increase productivity in the context of Industrie 4.0. Procedia. CIRP **19**, 51–56 (2014)
36. A. Albers, B. Gladysz, T. Pinner, V. Butenko, T. Stürmlinger, Procedure for defining the system of objectives in the initial phase of an industry 4.0 project focusing on intelligent quality control systems. Procedia. CIRP **52**, 262–267 (2016)
37. Bartodziej, Christoph Jan (2017), The Concept Industry 4.0 An Empirical Analysis of Technologies and Applications in Production Logistics
38. N.C. Batista, R. Melício, V.M.F. Mendes, Services enabler architecture for smart grid and smart living services providers under industry 4.0. Energ. Buildings **141**, 16–27 (2017)
39. M. Brettel, N. Friederichsen, M. Keller, M. Rosenberg, How virtualization, decentralization and network building change the manufacturing landscape: An industry 4.0 perspective. Int. J. Mech. Indus. Sci. Eng. **8**(1), 37–44 (2014)
40. Devezas, T,. Leitao, J., Sarygulov, A. (2017). Industry 4.0: Entrepreneurship and Structural Change in the New Digital Landscape

Impact of Internet of Things on Logistics Management: A Framework for Logistics Information System

Shivani Dubey and Vikas Singhal

1 Introduction

IoT handles business questions and difficult operations in an exciting new manner. Today, we see that IoT to overtake industry where technology pushes via mobile computing, information technology, big data analytics, and 5G networks from customers who are increasingly demanding IoT-based solutions. Many companies are using IoT to optimize different resources for improving logistics and growth of productivity. Tracking of products and information related to products at right time and right place can be beneficial for logistics to maximize the cost of storage. So, supply chain has gained an attention to be implemented with a new analysis and design of logistics for generating all operations with effectiveness and efficiency. IoT helps to implement each and every aspect of supply chain to improve logistics management, customization, automatic replenishment of products, demand management, and inventory management. IoT allows handling ubiquitous information about transformation of products via different types of devices and communication. This chapter describes the application of IoT in logistics management via supply chain to determine the tracking of information related to products delivery at right time and right place with a minimum cost from the supplier to customer. In this chapter, we present some IoT solutions for the logistics management of the complete supply chain, which creates the use of several existing IoT technologies like global positioning system (GPS), read-frequency identification (RFID), Wi-Fi, quick response (QR) codes, etc. in an efficient manner [1]. Most of the companies adopt the methods of logistics information system, which not only minimize the operating cost, but also increase the speed of response and implement the competitiveness of enterprises. There are still certain problems in the perspective of the

S. Dubey (✉) · V. Singhal
Department of Computer Science & Engineering, Lingaya's Vidyapeeth,
Faridabad, Haryana, India

© The Author(s), under exclusive license to Springer Nature Switzerland AG 2021
S. Awasthi et al. (eds.), *Artificial Intelligence for a Sustainable Industry 4.0*,
https://doi.org/10.1007/978-3-030-77070-9_2

operation of logistics information system like poor visual management, low auto-mation, frequent manual errors, higher data transfer cost, and higher response time for customer request. The research of IoT brings new opportunities to the innova-tion of logistics management. After mobile communication and the Internet, IoT is another revolutionary development of information industry. IoT includes a wide range of fields involved products and core technologies as well as collaboration and integration among several systems, networks, products, applications, and technolo-gies. Therefore, the research of IoT has been highlighted in the current scenario, and its related development and research have also been a major focus of a variety of countries.

IoT is a new business environment for logistics management where IoT devices enable identification of information, generating information, distribution of infor-mation, and location segment related to the product. Fully enabled IoT presents a virtual model for logistics in business systems, which is able to manage all the activities and processes at each level of logistics operations in real time based on providing the current state of logistics facility. Some time, logistics is based on communication and information technologies to support business connections and processes of customers via supply chain. It is best significant in logistics manage-ment to consider identification of information and communication related to the products. Identification technologies consider several operations and systems at each level of logistics such as packaging, vehicle routing, ports, GPS, RFID, and other systems used in logistics under the supply chain management [2]. In a logis-tics system, there are several models of connections using the Internet, which are related to multiple industries in supply chain. These models show the solutions based on IoT to lead global connection of all objects and users. According to Atzori et al. (2010) and Miorandi et al. (2012), the establishment of IoT and communica-tion network will enable providing a virtual model for business connections for all the users having data available in real time. In 2015, Macanlay et al. represented the solutions based on IoT with 75% in 2011 and 15% in 2012. So logistics depends upon the quality of logistics network and connections in supply chain for reliable and fast information. IoT has connectivity with identification technologies, software applications, large quantities of data, built-in intelligence, and decision-making at several control levels. The software systems used in logistics (logistics information system, warehouse system, time management system, operation, customer relation-ship management, and supply chain management) have maximum effects since high-quality information on the current state of objects on a network will be used [3].

2   IoT Emergence and the Internet

The Internet considers vast categories of protocols and applications built on inter-connected networks serving billions of customers around the world in a 24/7 man-ner. Kevin Ashton was accredited for using IoT for the first time in 1999 on supply chain management. He assumed that the way in which we live and interact within physical world that surrounds us needs serious reconsideration and the Internet by using smart devices. IoT produces various sensor-based and sharing-based services. The Internet of things (IoT) enables communication and interaction between different devices (equipment, products, and goods) as a part of a single system and an environment (Fig. 1).

The world has been striving toward this for decades. Nowadays, IoT becomes an integral part of people's lives. Due to new connected devices and the emergence of new technologies over different networks, a network's infrastructure has to be created surrounding the people, which resolves several issues solved independently. In future, the wall between the Internet and IoT will be disappeared completely in the universe [4]. IoT is an Internet connection concept with a range of electronic devices to people and products promoted using technologies like devices for storage, intelligent analytics, sensor devices, and decision-making analytics.

IoT is also called as interconnectivity of physical devices, buildings, vehicles, and other objects embedded with sensors, electronics, actuators, networks, and software, which enable these objects to exchange and collect the data [5]. These devices have a modified connection by using other advanced technologies like Bluetooth, Zigbee, RFID, Global System for Mobile Communications (GSM), QR scanner, and GPS considered as the Next Big Thing innovation [6] for predicting seven trillion wireless devices used in 2017. It is also expected to be expanding by 2020 to twenty-six billion and also help to monitor the human activities and machines [7].

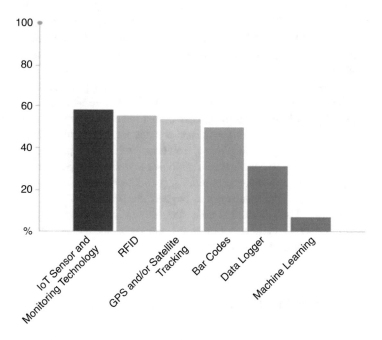

Fig. 1 Bar chart for IoT requirement in monitoring and RFID systems [3]

IoT is expanded in the corporate market and world by 21% to 19$ trillion dollars [8]. It has several advantages with a positive attention to logistics information system adopted by industries in their supply chain [9]. Some industries are still using traditional techniques that are so costly and unreliable for tracking the products in supply chain, and this has some issues like lacking in sharing the information in an efficient way by using different resources for taking the accurate decision [10]. It has further limited logistics management operations and its innovativeness. This chapter proposes a logistics information system framework based on IoT, which takes a support of different technologies, and presents conclusion at the last.

3 Interaction of IoT and Human

IoT interacts with participation and dissemination of information among the opportunistic communities formed on the basis of opportunistic an movement of person and device. Several personal devices like wearable phones, vehicles, and mobile phones can be the opportunistic forms of IoT with sensor modules and short-range communication. Several IoT devices have multiple characteristics to show unprecedented opportunities to understand the interaction of real-world entities with human. This interaction has been considered as a human environment, human–human interaction, and human object. To analyze the collection of these interaction data, data mining techniques and modern machine learning are used (Bin Guo, 2013). IoT technologies affected 120,000 jobs in countries by 2012. The Internet creates several businesses and fresh jobs that have functions never existed in the previous time. The new function of IoT supports cloud specialist, robot coordinator, pirate counter, agriculture technologist, etc. IoT is considered as a property of topology; a network is broadly categorized into two different ways: The first is an opportunistic connection or a dedicated connection and the second is an infrastructure-enabled connection. Opportunistic IoT addresses sharing and dissemination among opportunistic devices or a pair of devices based on opportunistic contact nature and movement. Several personal devices like wearable phones, vehicles, and mobile phones can be the opportunistic forms of IoT with sensor modules and short-range communication. IoT is equipped by three sensing capabilities: social awareness, user awareness, and ambient awareness. Social awareness is beyond the personal context, community levels, and groups. User awareness considers the ability to understand the behavioral, personal context and rules. Ambient awareness considers the status of information that includes traffic jams and space status.

4 Information Sharing in Logistics and Supply Chain

The performance of logistics service provider and supply chain depends upon the distribution of information among the suppliers or stakeholders; mostly information sharing minimizes the cost in whole supply chain. For example, effective information distribution plays an important role in supply chain management in addition to financial flow and products [12], and the integration of information distribution among the retailers and suppliers enables an accurate decision to provide an effective environment for improving customer service levels and to minimize the cost of logistics [13]. The extent of shared information can be classified as either tactical information (a low level of information sharing that can satisfy basic operation needs) or strategic information (a high level of information sharing that can satisfy strategic and competitive needs). Another view of information sharing, according to Seidmann and Sundararajan (1998), is that there are four levels of information sharing, i.e., order information, operational information, strategic information, and competitive information. However, of these four levels, order information is not considered as true information sharing. In a supply chain, dimensions of information sharing cover a range of aspects including information types, the quality of shared information, information sharing technology, and sharing implementation process (Marshall, 2015); benefits of information sharing (Kumar & Pugazhendhi, 2012, Yu, Yan, & Cheng, 2001); or security and information leakage (Kumar & Pugazhendhi, 2012, Tan, Wong, & Chung, 2016, Tran, Childerhouse, & Deakins, 2016) (Fig. 2).

There are various literature reviews based on IoT and supply chain. We summarize most relevant literature reviews and explain how our review is different from other existing reviews where Musa and Dabo (2016) focused on RIFD that is used in supply chain from 2000 to 2015. In 2017, Strozzi et al. presented a review on smart factories and also focused on manufacturing sector. The authors made restriction for their review to only those papers listed in Web of Science from 2007 to 2016. In 2017, Liao et al. represented Industry 4.0 with no specific attention on supply chain or applications of IoT that extend beyond Industry 4.0. In 2017, Naskar, Basu, and Sen described the study of RFID applications in supply chain [6]. According to Table 1, we defined some latest literature reviews on IoT in supply chain area. Our review is based on the role of IoT applications and its impact on logistics management in different sections of supply chain.

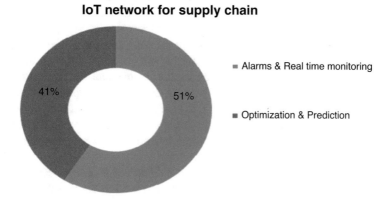

Fig. 2 IoT network for supply chain

Table 1 Existing literature reviews based on IoT and supply chain [7–11]

Authors (Year)	Development
Babamir, 2012	IoT has an advanced impact on routine of our life such as appliance maintenance and home management. It handles remote control, management of process quality, maintenance tools, and services for integrating supply chain.
Sivamani, 2014	Delivery of the product and real-time data at right place and right time allows the creation of new improvement and services of business processes.
Haass et al., 2015	The production of agricultural food and products to buy and sell from warehouse site to intelligent management of agricultural food and products improved their quality by using IoT.
Bi, Wang, Da Xu, 2016	Advanced IoT helps to monitor several wired sensor networks applications in transportation system, storage system, logistics, and automated work cells.
Yu Cui, 2018	A systematic research focused on the characteristics of supply chain system in IoT and analyzed the process of innovation of supply chain based on the architecture of cold chain tractability system.

5   IoT for Logistics Information System

Visibility is a big issue in logistics as there is a requirement of accurate information on timely shipment. In the current scenario, identification of transportation of what, to where, location, cost, all information required together from manual to automation with web sites, on paper, and phone calls is a much inefficient and time-consuming process. So there is a requirement of transparency into logistics operations in order to streamline the processes, minimize the errors, optimize cost, and eliminate redundant processes. To solve these types of issues in logistics, the companies are using logistics information system based on IoT framework as a solution that works as a centralized or distributed center where IoT provides end-to-end transparency within the complexity of logistics operations. These types of centers are using sensors, communication technologies, identification technologies, cloud storage, gateways, visualization technologies, business intelligence, business rules, etc. Logistics information system aims to collect and aggregate all orders, inventory, asset, shipment, etc. Information at a centralized or distributed center with an enterprise system to provide end-to-end transparency in real time to business users included suppliers, drivers, yard production, and warehouse. Technology provides energy into the business for increasing opportunities and leading multidimensional facts to understand demands from both the supplier and customer to understand logistics management. IoT helps to develop new digitization that gives the growth in logistics management. For handling business module easily with higher profits, advanced technology can be used such as sensor for dealing information like condition of road in transportation, location of the customer to deliver a product, and weather forecasting (Fig. 3).

There are several ways to fulfill customer satisfaction at a lower cost [13]. It is beneficial to several industries in small horizons as well as bigger rulers as it uses the network and intelligent devices such as personal computers or outsources the

	Identification	Position & Navigation	Condition Monitoring	Communication
GPS		■		
4G				■
Camera	■		■	
Wi-Fi		■		
Bluetoooth				■
RFID	■			
Gyro			■	
Temperature			■	
Humidity			■	
3D Barcode	■			

Fig. 3 IoT solution involved in logistics [12]

technical data and smartphone used directly for manufacturing level to collect data. Implementation of business can be done by IoT in logistics management with the following targets [14]:

1. Security Detection: To connect several applications, implement the control in warehouse system to track all items and alert to the business manager in case of missing of any product. For example, IoT application is connected to closed-circuit television (CCTV) cameras; then, application can allow the warehouse manager to close the door remotely, track the delivered products, and monitor logistics applications.
2. Employee Protection: Unreliable machines create a danger zone for employee's life. IoT effects on logistics with employee protection by detecting equipment with human interaction tools. IoT sensors are used to increase the response time, in case of any thing happened to an employee. A wearable device immediately detects a critical change and transfers data to dedicated platform that alerts and calls an ambulance and the manager.
3. Product Tracking: The primary objective of the manager is to increase transparency in delivering the products and to process with IoT implementation. This process develops confidence in the manager to check all the stages of the supply chain execution smoothly. It also catches the trust of customer with updating delivery status.
4. Advanced Analytics: IoT provides a view of big image to the manager to handle all the operations of logistics. Inventory monitoring and real-time delivery implement the quality of budget and planning.
5. Improving Quality of Delivery Process: IoT gives several directions to delivery management, where GPS sensors and RFID tags support the business manager for tracking shipment at its final stage. Logistics suppliers or managers can get real-time location data for ensuring the weather condition or environmental changes for the delivery of product.

6 Benefits of IoT for Logistics

IoT is the third eye of information technology, which includes several new and advanced changes and opportunities to logistics management. IoT provides technical support for advancing the visualization, realizing intelligent management, and the stability of logistics management. IoT officers have several benefits for logistics industries to access IoT-based logistics applications successfully. The most benefits of IoT are the following [15]:

1. Logistics Management Tools: Predictive analytics supports business managers in distribution and planning restocking. The owner of industry ensures the storage of product safely, saving time by being able to reduce human errors, and locating needed items, for example, Amazon proves this type of warehouse management and tracking.
2. Predictive Analysis Solutions: It helps business managers in predictive analytics to make accurate decision for logistics and warehouse management. These types of tools are utilized for detecting deficient equipment, replacement of equipment, and short delivery routes.
3. Automated Vehicles: Automated vehicles like truck, trolley, etc. will be new innovation in logistics. Logistics companies use self-driving vehicle and take benefits from logistics innovation to minimize the human error, control delivery process, and also maintain the intelligence of machine. The smart vehicle can select the most convenient path and also adjust the properties and temperature of vehicle according to the environment.
4. Drone Deliveries: Drones are used for automotive deliveries and to speed up the deliveries. In logistics, drone can be used to implement a navigation system within warehouse and also provides solutions for customer's delivery problems and instant in-store deliveries. This type of delivery reduces the amount of workforce and also cuts operational cost for keeping higher rates of customer satisfaction.

7 Innovation of Logistics in IoT Context

Innovation of logistics system under the IoT involves innovation of logistics processes and logistics operations. The development of IoT technology creates IoT-based logistics information system transformed significantly. The main function of traditional logistics information system involves only coordination, organizing, control, planning, etc. in the perspective of IoT; changes are needed in logistics to be adjusted. In general, logistics includes processing, sales, production, and procurement, which also need to be improved [16].

(a) Innovation in Coordination: Coordination is to promote the stability of operations by the cooperation and negotiation in company. The goal of logistics management is to increase the benefits of supply chain. IoT provides a technical support to increase the benefits of supply chain realizing intelligent management and visual management in logistics management for enabling to share information at each level of logistics operations.

(b) Innovation in Planning: Planning is an early stage of any organization. To achieve this target, logistics management using IoT is used for improving the level of visualization, stability, and transaction of logistics. The application of sensing devices like wireless sensors, the Internet, wired technology, and RFID tags can provide information about planning with better speed and precision to improve planning in an accurate way.

(c) Innovation in Organizing: The existing people-oriented system is replaced by an intelligent system for taking advantages at a lower cost and higher efficiency and extensive attention. The pattern of the new system transfers from the existing people-oriented pattern, which is oriented by manual machine and intelligent system. In the context of IoT, the exertion of organizing function involves the relationship among people and also includes the cooperation between machines and people to coordinate with machines.

(d) Innovations in Controlling: The controlling function makes the comparison between the intended objectives and implementation outputs and then presents the feedback results corresponding to that particular output. In the emergency task, IoT handles the different kinds of instructions and different parameters related to the different sensors.

(e) Innovation in Marketing: Marketing and sales are adopting IoT technology for providing unusual and convenient services. IoT application in marketing also handles various methods; e.g., cashier can use smart cash register for checking out and scanning the products and display the total price in the shopping cart, and RFID can quickly read and display price. In payment methods, customer can select the bank card payment, Paytm payment, cash payment, mobile payment, fingerprint payment, etc.

8 IoT Framework for Logistics Information System

In growth of IoT, the rapid rates of devices and sensors are becoming more common for communicating information among them. In business where information like temperature, position, and other properties is required to be identified, IoT can help to produce accurate tampering data. Logistics information system is a novel way to consider the business automatically among the business partners without the requirement of server and centralization, and all the users can access the same data at a time from different locations [17]. How logistics management can take benefits when data are distributed through the IoT? IoT-enabled package transfers the required information related to logistics through several carriers and specifies the condition of logistics shipment from one place to another place, and all the suppliers and users follow the contract. Using IoT to logistics will allow to access the same information without requiring central control for all logistics partners. Logistics information system provides a mechanism to proceed to be affirmed by unreliable users. It has a distributed environment to secure and transfer the information at different levels of supply chain. There are some issues that have been observed in the combination of IoT and logistics information system like continuity in information flow, effective exchange of information among several logistics partners and users, effective access of the link between a physical link and information flow at all the logistics stages, and fraud and violation detection.

9   IoT for Logistics Information System in Future

The logistics industry is leading the direction for the explosion of connect systems. IoT is a major trend with significant implications for global economy. IoT will be a game changer. According to the survey of Forrester Research, 80% of global industries agree to adopt IoT-based solution for their organizations. At the level of global scale, a new digital infrastructure must be required to enable digital and physical services that work as smart services. These infrastructures create a critical role for economy transformation in society. Several companies are using multiple networks and sensors to distribute information by using multiple devices for more than 25 years, but in the current scenario, everything is on cloud, which enables the processing of large amount of data and storage with advanced software to extract and manage useful information quickly at a minimum cost [18]. Sensors, communications, and different microprocessors are adopted rapidly and connected with intelligent devices. IoT is a reality in living devices; we can see IoT applications in robotics, with multiple sensors, transportations, and communication devices, as well as in manufacturing of the products also. IoT includes an integral space in the growth and improvement of logistics industry, which provides smart transportation system, information distribution, warehouse management, and logistics operations for implementing supply chain using intelligent and smart devices [19]. Below are directions in which IoT technology helps in logistics information system:

(a) Safety of Information: Information tracking is a crucial aspect of inventory and warehouse management; mostly companies face challenges to ensure the safety of employees and staff using IoT for minimizing the chances of accidents. Companies are using different kinds of sensors to monitor the operations of vehicles and also to minimize the risk of work-related failures. By using equipment based on IoT, an alarming system can detect and alert the people to update the safety of products and employees also.

(b) Security of Inventory and Warehouse Management: Security is the most important part of inventory and warehouse management. IoT devices help to keep assets and information secure by preventing theft. IoT technology ensures the security of inventory and warehouse management by using sensors and video surveillance cameras for detecting theft in both management systems. Remote sensing and alarming system help to monitor video data at any given time among the different locations.

(c) Performance Improvement: IoT provides the facility of interconnectivity of devices that allows better management process to logistics information system, which is essential for business advancement. Logistics companies use IoT that controls the full view of business, and IoT technology allows people to have a view of the entire process of logistics operations by using resources like sensors, communication, and identification technologies.

(d) Minimizing Risk and Optimization of Logistics Efficiency: Tracking of a product in real time is not just as the transmission from one place to another place, but is each single moment until it reaches a customer's location. In tracking the vehicle autonomously, big data plays a very important role with IoT to increase the efficiency and minimize the cost throughout the information distribution in logistics information system by the space optimization, smart dispatching, and route optimization among other benefits.

10   Conclusion

IoT provides logistics management to take the facility of logistics information system with transparency and control that have not seen in the traditional environment of logistics management. Only IoT provides device-to-device communication for real-time data sharing with advance technologies that improve streamlined delivery of products and decision-making for all the logistics operations. IoT brings visibility, tracing of shipment, improved productivity, safety of products, and tracking of products into the logistics operations and functions. Logistics providers and manufacturers can utilize the cloud services, sensor technology, and analytical capacity by using the framework of logistics information system based on IoT. IoT also minimizes the higher cost of communication cloud, data visualization technologies, and expensiveness of sensors. IoT is a key technology for logistics information system to improve and control the performance of logistics management via supply chain that contributes to the success of any organization. In this chapter, a framework of logistics information system based on IoT is provided, which brings more benefits to inventory and warehouse management for overall performance improvement into the logistics operations and functions. This chapter is also presented that the framework of logistics information system based on IoT can improve the inventory shortage, efficiency, order delivery, the profits, and warehouse management. IoT is not just a technology, but it is also a combination of many technologies that had been invented in a very effective way to enhance the logistics information system for better work field and cost reduction. The autonomous industry uses IoT as a core business into a vehicle sector for transportation like airway, railway, maritime, and roadway, which will be more secure and effective. In respect to connect all the devices with different vehicles for tracking the location of products, we have to use more sophisticated programs with advanced software to keep information in a safe mode.

References

1. C. Caballero-Gil, J. Molina-Gil, P. Caballero-Gil, *Alexis Quesada-Arencibia, IoT Application in the Supply Chain Logistics, Computer Aided Systems Theory-EUROCAST* (2013), pp. 55–62
2. E. Borgia, The internet of things vision: Key features, applications and open issues. Comput. Commun. **54**(1), 1–31 (2014)
3. Macanlay J, Buckalew L, Chung G, Internet of Things in Logistics, A collaboration report by DHL and Cisco on implications and use cases for the logistics industry, 2015
4. M.A. Rahman, A.T. Asyhari, The emergence of Internet of Things (IoT): connecting anything, anywhere. Journal of Computers/MDPI (2019)
5. A.V.K. Prasa, *Exploring the Convergence of Big Data and the Internet of Thing* (IGI Global, 2017)
6. R.M. Weber, Internet of Things become next big thing. J. Financ. Serv. Profession. **70**(6), 43–46 (2016)
7. J. Rivera, L.Goasduff, Gartner Says a Thirty Fold Increase in Internet Connected Physical Devices by 2020 Will Significantly After How the Supply Chain Operates, 04 May, 2017
8. O. Kharif, Cisco CEO Pegs Internet of Things as $19 Trillion market
9. P.R. Nair, V. Raju, S. Anbuudayashankar, Overview of information technology tools for supply chain management. CSI Communications **33**(9), 20–27 (2016)
10. Z. Michaelides, Big Data for Logistics and Supply Chain Management, Production and Operation Management Society (POMS) Conference Proceedings in Orlando, Florida, 2016
11. The Internet of Things (IoT) in Supply Chain and Logistics, Research findings, 2016, timan-box.com
12. https://in.nec.com/en_IN/pdf/Leveraging_IoT_for_Logistics-1.pdf
13. M.A. Basset, M. Gunasekaran, M. Mohamed, *Internet of Things (IoT) and Its Impact on Supply Chain: A Framework for Building Smart, Secure and Efficient Systems* (Future Generation Computer Systems, 2018, April)
14. S.M. Babamir, M2M architecture: Can it realize ubiquitous computing in daily life? KSII Transactions on Internet and Information Systems **6**(2), 566–579 (2012)
15. Sivamani, S., Kwak, K, Cho Y, A study on intelligent user centric logistics service model using ontology, J. Appl. Math., v 2014, 2014
16. R. Haass et al., Reducing food losses and carbon emission by using autonomous control-a simulation study of the intelligent container. Int. J. Prod. Econ. **164**, 400–408 (2015)
17. Z. Bi, Z.G. Wang, L. Da Xu, A visualization platform for internet of things in manufacturing applications. Internet Res. **26**(2), 377–401 (2016)
18. C. Yu, *Supply Chain Innovation with IoT, Multi Criteria Methods and Techniques Applied to Supply Chain Management* (Intechopen, 2018)
19. M.B. Daya, E. Hassini, Z. Bahroun, Internet of Things and supply chain management: a literature review. Int. J. Prod. Res. **57** (2019)

Green Internet of Things (G-IoT): An Exposition for Sustainable Technological Development

Prerna Agarwal, Pranav Shrivastava, and Satya Prakash Singh

1 Overview

The use of computationally advanced devices, such as smartphones, is increasingly growing. Over the last decade, the alarming levels of power consumption have been triggered by large numbers of users and their various devices. According to a conservative estimate, connected devices will number about 50 billion by 2020 [1] and 100 billion by 2030 [2]. It is projected that cellular networks will generate more than 345 million tons of carbon dioxide (CO_2) by 2020 and that this trend will continue in the coming years [3]. Full pollution forecasts of 2020 are given in Fettweis et al. [4]. Green or sustainable technology has emerged as an important area of inquiry in technological development as a result of these significant emissions and financial and medical issues. Furthermore, existing computer battery technology is another big problem that propels green technology forward [5].

According to experts, the fifth generation of wireless communications (5G) will be completely deployed by mid-2020, capable of handling approximately a thousand times more cellular data in comparison with the existing cellular networks [6]. As depicted in Fig. 1, there are five strong 5G connectivity technologies. The IoT's capabilities allow billions of users to stay connected. Device-to-Device (D2D) networking enhances the speed and efficiency of user-to-user communication by decreasing latency, and spectrum sharing (SS) helps to mitigate the effects of

P. Agarwal
JEMETC, Greater Noida, UP, India

P. Shrivastava (✉)
G.L.B.I.T.M, Greater Noida, UP, India
e-mail: pranav.shrivastava@glbitm.ac.in

S. P. Singh
BITS, Mesra, Ranchi, India
e-mail: sp.singh@bitmesra.ac.in

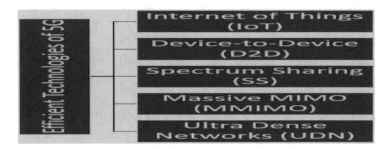

Fig. 1 Efficient technologies of 5G

inefficient spectrum use. Furthermore, ultra-dense networks (UDNs) entail a large-scale small-cell distribution in high-traffic areas. In contrast, the large multiple-input and multiple-output (MIMO) allows for hundreds of antennas, resulting in an augmented data rate. These game-changing technologies would make it possible to control the energy use of future 5G networks, lowering CO_2 emissions. The main aim of this chapter is to provide a review of green IoT concepts, implementations, inventions, and challenges.

2   Internet of Things

In 1999, one associate of Radio Frequency Identification Development Society (RFID DevCom) hypothesized the Internet of Things (Fig. 2). The growth in the usage of mobile devices reinforced by ubiquitous communication, snowballing computation power, and data analytics has rendered it additionally germane to the practical world. Visualize an integrated scenario where billions of devices will be proficient at sensing, communicating, and sharing information. Such integrated structures comprise data collection, processing, and usage to enable systematic actions, endowing strategy, control, and administration in conjunction with a variety of information. This is the Internet of Things (IoT) culture [7].

The public notion of IOT is:

> Internet of Things (IOT) is a network of physical objects. The Web is not only a network of computers, but it has grown into a network of devices of all shapes and sizes, cars, tablets, home appliances, toys, cameras, medical devices and industrial systems, livestock, humans, houses, all linked ,all interacting and sharing information on the basis of protocols to accomplish smart reorganizations, mapping, tracking, surveillance. [1, 2].

The Internet of Things is the three-dimensional Internet; we describe IoT as follows into three categories: users to users, users to devices, and devices to devices. IoT is a perception and a benchmark that anticipates inclusiveness in the world of countless devices that are able to interact among themselves, through wired and wireless links and special addressing schemes, cooperate with other devices to conceive novel applications/services, and achieve shared purposes. The research and

Fig. 2 Internet of Things [10]

development intricacies in expounding a sustainable environment in this sense are colossal, a prospect in which physical, digital, and virtual congregate collectively to fabricate smart worlds [8, 9].

IoT relates to the common outline of things, in particular general objects, which are visible, identifiable, traceable, addressable by data-sensing devices, and/or manageable via the Internet, regardless of connectivity media (whether via RFID, wireless local area network [WLAN], wide area networks [WANs], or further channels). Everyday objects comprise of electronic devices we use in our routine procedures, advanced technical products like equipment and vehicles, and in addition objects we don't usually perceive to be electronic, like clothes, food, animals, and so on [8, 9].

IoT is a new cyber movement where devices are identifiable and gain wisdom as a result of taking or permitting circumstantial judgments, as they are capable of conveying knowledge regarding them. Any device is able to access data aggregated by other entities, or they may become elements of structured systems. This revolution corresponds to cloud computing advancements and the Internet's move to Internet Protocol version 6 (IPv6) with limitless addressing power [8, 9]. The aim of the Internet of Things is to provide the ability to link objects anywhere, wherever, with anyone and everyone preferably exploiting whichever network or whichever device as depicted in Fig. 3.

2.1   The Technology Stack of IoT

If anyone wants to navigate the IoT engineering labyrinth, due to the complexity and sheer numbers of technology solutions that accompany it, it can be a tough task. However, we can break down the stack of IoT technologies into four simple levels of technology concerned with driving IoT.

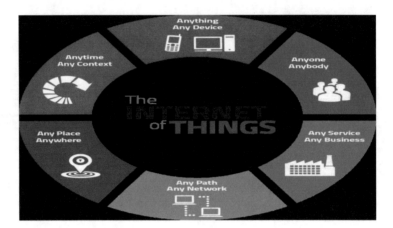

Fig. 3 Internet of Things [11]

(a) Device Hardware: Devices are artifacts that are essentially the "things" within the IoT. They act like an intermediary between the digital and physical infrastructures, assuming diverse dimensions and complexities based upon their role as per the requirement, contained by the particular IoT implementation. Using miniscule spy cams or large structural devices, nearly any object (humans or animals) may be converted as a connected device via installing required instrumentation (through attaching actuators or sensors together accompanied by the correct software) to calculate and accumulate the needed data. Actuators, sensors, or any supplementary telemetry devices might, of course, also perform as stand-alone smart devices. Individual drawback found is genuine case for IoT usage and hardware specifications (dimensions, simplicity of installation and maintenance, durability, helpful service life, and cost-effectiveness).

(b) Device Software: Software is accountable for realizing the Internet connectivity, data collection, system deployment, and data monitoring within the IoT network. Additionally, it also provides consumers with the computer functionality to imagine data and communicate with the IoT framework with application-level capabilities.

(c) Communications: Even as it is possible for communication frameworks to be strongly coupled with computer software and hardware, treating them as a separate layer is important. Communication framework contains all physical communication mechanisms (cellular, cable, and local area network [LAN]) and common protocols used in different IoT environments (Zigbee, Z-Wave, Thread, Message Queuing Telemetry Transport [MQTT], Lightweight mobile-to-mobile [LwM2M]). Selecting right connectivity approach is an essential component in constructing a stack of IoT technologies. The selected technologies would decide not only cloud communication protocols, but also systems handling and interaction with third parties' apps.

(d) Platform: As suggested, computer can "feel" what's going on around it thanks to the smart hardware and the enabled apps and relay that to the consumer through a different communication channel. The IoT framework is the location to capture, organize, store, interpret, and display all of these details in a comprehensible and intelligible manner. This renders such a system particularly beneficial; however, it is not only its data collection and processing capability, but its capacity for analyzing and finding useful information from the data portions generated as a result of the devices' interaction by the communication layer. However, there is rather a range of IoT solutions available in marketplace, having preference based on the particular IoT project specifications and considerations such as stack design and IoT technologies, usability, configuration properties, common protocols, device agnosticism, protection, and cost-effectiveness. In addition, on-premise or cloud-based systems can also be built, e.g., Coiote IoT Device Management is capable of deployment both on-site and within cloud.

2.2 IoT Stack: Connectivity Options

The problem related to shortage of connectivity solutions does not hinder practical purposes of the IoT technologies. The question is "Can communications solution offer different service-enabling scenarios based upon requirements of provided IoT use-case while providing balance between energy demand, coverage, and bandwidth?" For example, suppose if anyone is creating smart home, he might wish for having internal temperature sensors and heating controller assimilated with smartphones so he may supervise the temperature of every room remotely and adjusts them to the current needs. In likewise cases, recommended solution would be the Thread, an IPv6 networking protocol, specifically premeditated for home-automation system.

This complexity and diversity related to networking norms and protocols raises another issue regarding actual requirement to develop innovative implementations even a number of Internet protocols already exist from decades. This is happening because accessible communication protocols, such as Transmission Control Protocol (TCP)/IP, are inadequately powerful and require more energy for working proficiently in evolving IoT implementations. Each segment will provide a brief overview of the main alternate Internet protocols dedicated specifically to IoT devices. The summary discusses the most common IoT network applications, broken down by the spectrum of radio frequencies reached through the following solutions: short-range solutions, medium-range solutions, and long-range solutions.

Short-Range IoT Network Solutions:

(a) **Bluetooth** It is a proven short-range communication protocol and is regarded as a main remedy, especially in support of prospects of wearable technology devices, particularly due to its universal inclusion in smartphones. The Bluetooth Low Energy (BLE) protocol is deliberated with practical and reduced energy consumption and necessitates incredibly minute power from device. However, there exists a downside: This protocol isn't the best practical solution while communicating bigger data quantities regularly.

(b) **Radio Frequency Identification (RFID)** As it is a pioneering IoT technology, it offers promising IoT solutions, particularly for logistics and supply chain management (SCM) that involve capability to determine entity location within buildings. Clearly, RFID prospects surpasses uncomplicated localization services, having possible implementations varying from managing patients for augmenting competence of healthcare services to furnishing real-time inventory information to mitigate out-of-stock conditions for outlets.

Medium-Range Solutions:

(a) **Wi-Fi** It is established on IEEE 802.11 and continues to be the most extensively used and commonly recognized protocol for wireless communication. The extensive use throughout the IoT environment is primarily restricted due to elevated energy consumption because of the requirement to maintain signal strength and swift data transmission in favor of enhanced efficiency and connectivity. Information technology (IT) offers an extensive range of ground for surprising numbers of IoT applications; nevertheless, it in addition necessitates management and regulation in marketing to deliver revenues to both consumers and service providers. Linkify is one good example of a Wi-Fi management platform that provides a value-added service that empowers public Wi-Fi access points.

(b) **Zigbee** It sees its largest use in household electronics, traffic management systems, and the machinery industry. It is developed on the standard IEEE 802.15.4. It facilitates low rates for data exchange and low power service, safety, and consistency.

(c) **Thread** It is premeditated exclusively for smart home items and uses IPv6 networking to allow connected devices for interacting among themselves, access cloud services, or provide consumer interaction through Mobile Thread applications. Thread's analysts have suggested that with the competitiveness of the industry yet another protocol for communication is contributing to supplementary disintegration inside IoT stack.

Long-Range Wide Area Networks (WAN) Solutions:

(a) **Narrow Band-Internet of Things (NB-IoT)** It is the latest standard of radio technology that guarantees exceptionally low energy utilization (battery power of ten years) and supplies connectivity with an approximate signal strength of 23 dB smaller than in the second generation of wireless communications (2G). It also utilizes an accessible communication framework that imparts global coverage of Long-Term Evolution (LTE) networks and in addition boosts signal efficiency. The mentioned advantage allows NB-IoT to be implemented instead of solutions that require local network building, such as Sigfox, LoRa, etc.

(b) **LoRaWAN** It is a long-range, low-power WAN protocol augmented in favor of low energy usage and supporting millions of devices in large networks. It is premeditated to provide low-power WANs for supporting inexpensive, portable, and safe two-way communication in IoT, smart cities, and industrialized applications for wide area network (WAN) applications.

(c) **LTE Cat-M1** LTE Cat-M1 is a standard for low-power wide-area (LPWA) connectivity that connects M2M and IoT devices with medium data requirements. This facilitates extended battery life cycles and has an increased range of building relative to wireless systems such as 2G, third generation of wireless communications (3G), or LTE Cat-M1. Because Cat-M1 is compliant with the current LTE network, it does not allow the carriers to build new networks for deployment.

(d) **Sigfox** It provides an efficient networking approach for low-power M2M applications that require short data transmission rates for which the cellular range is extremely costly and power-hungry, and the Wi-Fi range is too limited. It uses Ultra Narrowband (UNB), a platform that helps to accommodate low data transmission speeds ranging between ten and thousand bits per second. It offers a reliable, power-efficient, and flexible network capable of sustaining connectivity between thousands of battery-operated devices over several square kilometers of land. Sigfox is ideal for numerous M2M applications including smart meters, smart patient alarms, street lighting, environmental sensors, and safety devices. It is currently employed in an increasing number of IoT technology solutions, such as the Coiote IoT Software Orchestration by AVSystem.

2.3 Empowering Technology for IoT

IoT is a universal information infrastructure that permits highly developed services through interconnecting (virtual and physical) objects/devices centered upon accessible and developing ICTs. With IoT, the network is applied to all the objects/devices surrounding us via Web. IoT is so much more than just machine-to-machine connectivity, cellular sensor networks, sensor networks, Global System for Mobile Communications (GSM), General Packet Radio Service (GPRS), RFID, Wi-Fi, global positioning system (GPS), microcontroller, microprocessor, etc. These are the underlying technologies that make "Internet of Things" feasible. The enabling technologies for Internet of Things can be divided into three categories [8]:

1. Technologies enabling "things" to collect relative and appropriate data.
2. Technologies enabling "things" to handle contextual information.
3. Technologies to enhance privacy and security.

The first two definitions can be interpreted together as conceptual building blocks that allow the integration of "intelligence" into "things," which are indeed the characteristics that separate the IoT from the ordinary Internet. Last category is not a requirement related to functionality; instead, it is an imperative and a genuine prerequisite devoid of which the IoT penetration might be unrelentingly diminished [9] ← [2]. IoT is often promoted as a stand-alone technology; on the contrary, it is a combination of several software and hardware innovations. IoT offers resolutions centered on top of IT integration that purport to have software and hardware exploited for storing, retrieving, and processing information and networking that comprises electronic systems intended for group or individual interaction.

The diverse combinations offered by current networking technologies necessitate modification for becoming appropriate for requirements of IoT applications like efficient power consumption, reliability, security, and speed. Pertaining to this, it's likely that diversification may remain limited, toward a range of achievable networking solutions that deal with requirements of IoT implementations, are embraced by the industry, already proved to be serviceable, and backed by a broad development partnership. Definitions of such specifications cover wired and wireless systems such as GPRS, GSM, Bluetooth, Wi-Fi, and Zigbee [8, 9].

2.4 IoT Features

The following features are IoT basic qualities [9, 12]:

(a) **Interconnectivity:** Any device capable of integrating with global communication network will be able to implement IoT.
(b) **Things-related services**: IoT is competent to deliver services related to objects/devices, contained by restraint of things, like confidentiality and functional continuity of material things and their abstract corresponding items. For providing object-related services within the constraint of things, both the knowledge environment and physical world technology will be evolving.
(c) **Heterogeneity:** Amalgamation from various networks and platforms leads to heterogeneity in IoT systems. These systems might correspond with other devices or application platforms, within several networks.
(d) **Dynamic changes:** System state varies continuously, e.g., powering up and down, linking or/and disconnecting, and system dimensions like position and velocity. In fact, the device's number might alter vigorously.
(e) **Enormous scale:** The quantity of devices that need supervising and that contact all others will be greater than the devices connected to the current Internet. The control of the produced data and their analysis for the purposes of implementation will be even more important. It includes data semantics, as well as secure data handling.
(f) **Safety:** As IoT provides benefits, safety as an imperative issue must not be ignored. IoT developers and users both have to plan for protection. It requires the protection of our personal data and our physical well-being. Safeguarding data and networks necessitates developing a scalable security model.
(g) **Connectivity:** Connectivity enables compatibility and accessibility of networks. Compatibility provides the ability to generate and consume information while accessibility is the ability to access a network.

2.5 Architecture of IoT

There is no overarching standard for IoT design, which is universally agreed. Several scholars have suggested various architectures. The primitive architecture is an architecture with three layers [13–15] as depicted in Fig. 4. It was instituted at the early stages of research in this domain. It comprises of three layers, namely the layers of application, network, and perception:

(a) The *perception layer* comprises sensors to sense and collect environmental data. This layer detects certain physical parameters in the world or recognizes certain smart objects.
(b) It is the duty of the *network layer* to link up additional servers, network devices, and intelligent things. Its characteristics are in addition utilized to transmit and process sensor data.
(c) It is the responsibility of the *application layer* to provide the user with specific application services. This describes various applications in which, e.g., smart houses, smart cities, etc. can be implemented on the Internet of Things.

The design with this trio of layers expresses primary notion of IoT, but it's not appropriate for IoT research, as that is often based on finer aspects of the IoT. This led to a more complex architecture projected in literature. One of the designs is of five layers that also comprise the business and processing layers [13–16]. The five levels are business, application, processing, transport, and perception layers (Fig. 3). The application and perception layer have the same function as in the three-layered architecture. The remaining layers are illustrated as follows:

(i) **Transport layer** transfers data from sensors to and fro between the processing layer and perception layer through networks such as Near-Field Communication (NFC), RFID, Bluetooth, 3G, LAN, wireless, etc.
(ii) **Processing layer** is furthermore classified as having the status of *middleware*. This collects, analyzes, and manages huge data volumes coming through the transport layer. Middleware is able to handle the lower layers and support them with varied services. For processing big data, it employs numerous technologies, such as cloud computing, databases, etc.

Fig. 4 Architecture of IoT (A: three layers) (B: five layers)

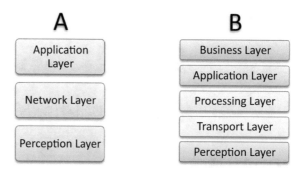

(iii) **Business layer** controls complete IoT network, incorporating apps, market and benefit structures, and the consumers' privacy.

The design anticipated by Wang [17] was influenced as a result of the study of information processing within human brain. Human acumen and capability to perceive, sound, recall, take judgment, and adjust to their environment have a direct influence on their work. Their model comprises three sections:

(a) A unit for computing and data management or the data center, akin to human brain.
(b) The elements and sensors in the networking resembling the nerve network.
(c) The remote data processing nodes and smart gateways network similar to the spinal cord.

2.6 Ecosystem of IoT

IoT can be thought of as an ecosystem that provides data transfer network interconnected with cloud computing and big data to offer intelligence for recognizing behaviors and explaining reactions established on data summarized through smart devices/objects accessible in up-and-coming smarter cities devoid of user-to-user communications. Figure 5 illustrates the IoT network design, wherein information obtained through smarter cities is integrated into the cloud servers. The contact between the cloud and users, which are starting to be more involved (presumed), is provided through this flow [18]. Centralizing data on behalf of every and all sensors and entities are responsibility of cloud computing system. By the formation of an omnipresent network, it also helps in connecting and communicating. For reaching a consensus that permits the determination of patterns of human dynamics, the cloud allows for the incorporation of big data analysis. Eventually, human

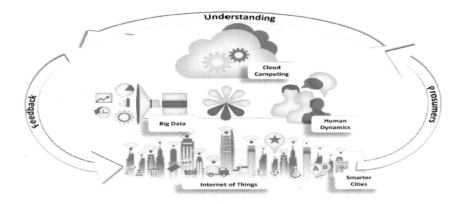

Fig. 5 IoT ecosystem

evolutionary structure includes input channels and tools to encourage behavioral change [18, 19].

3   Green IoT

Internet of Things (IoT) includes massive usage of the development network and interconnected devices. Therefore, the resources needed to implement the entire element set of networks and their operational energy consumption need to be kept minimal. Power usage is a crucial and an imperative criterion to achieve sustainability in green IoT and smart planet deployment. To have a healthy digital environment and minimal CO_2 pollution and adverse effects on environment, IoT requires improving efficiency of each and every system component, like sensors, computers, software, and utilities. Illustration 6 shows the life process of green IoT taking in account green design, green production, green utilization, and eventually disposal and recycling to have minimum or no environmental impact [20] (Fig. 6).

3.1   Key Technologies of Green IoT

The key contribution for making efficient Green IoT is depicted in Fig. 7:

(i) **Green Tags Networks.**
    The tags used to identify individual items should be environment-friendly, maintenance-free, and sustainable.

Fig. 6 Life process of green IoT

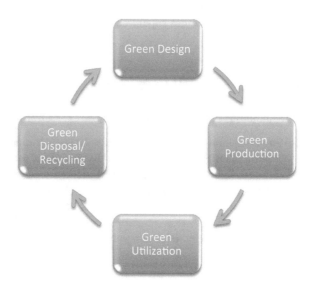

Fig. 7 Key technologies
for green IoT

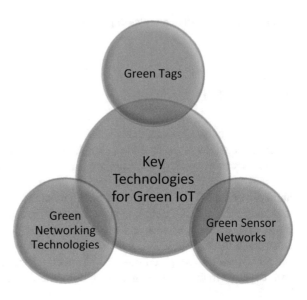

(ii) **Green Internet Technologies.**

For green Internet technology, hardware and software should be taken into consideration where hardware solution manufactures devices that consume less energy without a reduction of the performance. On the other hand, the software solutions offer efficient designs that consume less energy by minimum utilization of the resources.

(iii) **Green Sensing.**

A green wireless sensor network (WSN) is another key technology to enable green IoT. Wireless sensor network (WSN) contains a tremendous number of sensor nodes with limited power and storage capacity. To achieve green WSN, different techniques should be considered.

3.2 Application of Green IoT

The numerous green IoT implementations are as follows:

(a) **Smart Home**: A G-IoT allows smartphone or computer to manage illumination, cooking, and electronic devices to be controlled remotely. Smart beds, garbage disposal, lighting fixtures, machines, clothes depending on your preference, temperature, doors, and walls can enable unlimited levels of sunshine, hot or cold. Digital soundproof rooms and doors are few more implementations for a smart home. Hidden electronic microphones, sensors, and computers inside house are soundproof energy fields that you can walk through. Central computer acknowledges voice commands and differentiates inhabitants for custom answers and acts. TV, smart screens and mobile phones etc., combine and act as a single unit.

(b) **Industrial Automation:** Industries have been automated with machines that, without or with little manual intervention, allow for fully automated tasks. An Internet-based automation network for the market that enables the operation of manufacturing equipment by a single industry operator.

(c) **Smart Healthcare:** IoT intends to remodel the healthcare industry by creating new and sophisticated devices connected to the Internet that produce critical real-time data. This helps achieve three main results of effective health services–enhanced accessibility, improved quality, and lowered costs.

(d) **Smart Grid:** In conjunction with IoT, a smart grid concerns efficiency and equilibrium. It's about constantly modifying and readjusting to produce electricity optimally at the lowest possible expense and the maximum permissible efficiency.

4   IoT Big 7: IoT—Challenges and Problems

IoT permits devices/things/objects for communicating data and provides these services via Internet protocols. Together with several heterogeneous artifacts, complexity of the interconnected world demands resolution for seven most important challenges: interoperability, scalability, reliability, efficiency, availability, storage, and security. Gartner [21] speculated that the IoT might comprise 26 billion devices by 2020, creating new problems in all areas pertaining to data centers. IoT technologies often necessitate real-time big data processing, resulting in an escalation of data center burden and creating new protection, efficiency, and data analytics challenges [22]. Below is the summary of the difficulties.

4.1   IoT Scalability

A scalable IoT system should be capable of connecting new objects/devices, new members, and new diagnostic abilities and technologies capable of imparting continuing support. In case new users or/and devices wish to join the infrastructure or network, IoT scalability ought to regard prospects of offering high quality of service (QoS) (low latency, analytics, etc.). IoT scalability poses a key challenge for our society today, owing to quick escalation of connected devices, numerous technical modifications, and humongous IoT interactions, on top of exploiting cloud computing to attend to challenges posed by the Internet of Things as the growing demand for services.

4.2   Interoperability of IoT

Despite the complexity of multiple IoT-integrated systems, interoperability is an
extra main test that IoT will countenance in order to deliver services effectively and
to share data. As per a McKinsey's report [23], despite the rapid development of IoT
systems, their interoperability is expensive, in the range of four trillion dollars or
40% of the total operational value by 2015. Effective IoT interoperability includes
specifications that allow different types of devices to connect, function, and integrate. Acquisition, data sharing, processing, and the use of network data have been
very problematic so far. Interoperability might be evaluated in the context of considerations like information processing degree (organizational, technical, syntactic,
and semantic) or at the time of achieving interoperability (dynamic and static)
[24, 25].

In the analysis of information, there are four directions:

- Organizational (total information relating to organization).
- Technical (means of data representation on physical media).
- Syntactic (syntactic information representation in data gathering, communica-
  tion, recording, and processing).
- Semantic (the meaning of the data).

The analysis of interoperability is done at the level of software/hardware, platforms, and systems, which facilitates device-to-device (D2D) communication at the
technical level. These investigative analyses concentrate on communication infrastructure and protocols. Starting on the four interoperability levels, Fig. 8 identifies
the interoperability problems that are specific to each level (processes, details, documents, and devices). The IoT constitutes a broad array of applications that pose
problems in terms of static interoperability (conformity of requests accomplish-
ment). Partial noninteroperability is acknowledged (e.g., for some protocols), in
case they are addressed in sight (dynamic interoperability). Smart middleware and

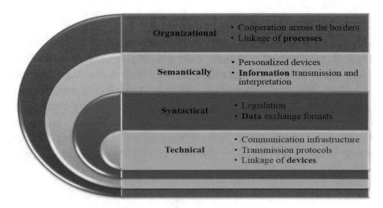

Fig. 8 Challenges by the level of interoperability in IoT

gateways from heterogeneous and dynamic IoT environments have demonstrated complex interoperability capabilities [24].

4.3　IoT Reliability

For any device, reliability is characterized in the function of system's capability for reliably executing any mission/request/task exclusive of fault/failure, a concept that moreover favors IoT context. As IoT constitutes large data volumes where information desires proper collection, storage, and control, modern technologies are required to efficiently and effectively disseminate and process information. Many IoT implementations have to run for limited duration that needs a long-tenure investment. Underneath these scenarios, network must be adequately adjustable to adapt to environmental circumstances or necessary improvements in components of the network. To ensure better reliability, attempts for standardization are crucial. Kempf et al. (2011) outlined four main areas of investigation and analysis that standardization raises: reliability in device architecture design, reliability in system development, load bearing of mobile network sensor gateway connectivity, and reliability of communication [26].

Device reliability becomes a real challenge due to the quantity and variety of interconnected objects. The network should consistently provide right, constant, and efficient data access and good data processing depending on the type and function of a computer, even in an instance of missing control, low or no Wi-Fi signal, if an access point or a server breaks down. Network operators are challenged by numerous umpteen users, having constant Internet connection in support of an assortment of online services, conjointly with IoT network having a large number of connected devices, because they must assure a continuous and reliable high-speed access to network. IoT's progress is highly dependent on network efficiency; therefore, the entire IoT infrastructure as well as network providers at each level must be evaluated robustly. An additional challenge in securing an uninterrupted broad access at the level of complete IoT infrastructure is multiple types of networks or network operators providing several services. The provided service reliability is associated by means of assuring superior availability, compilation, and storing and analysis of vast data volumes obtained through IoT devices without mistake. As per these deliberations, IoT interoperability seems strongly linked to reliability.

4.4　IoT Efficiency

Another challenge is ensuring efficiency of IoT infrastructure in a connected environment. Network must be capable of:

- Supporting multiple real-time analysis of vast data volumes existing within network.
- Fulfilling the variety of demands for data processing.
- Processing data as soon as they enter independent storage position (application decentralization).

An additional challenge is ensuring advanced data analysis and processing in support of view to intelligent devices with embedded processes of machine learning. For this reason, explicit constituents like artificial intelligence, intelligent agents, neural networks, etc. should be added to the IoT structure. From the same performance viewpoint, we should look at the IoT network's economic angle and calculate the impact of its activity against the effort put into its operation in a well-defined sense. In fact, it transforms into an aspect of the higher-order feedback loop demonstrating a systematic activity in relation to its setting by giving the IoT a cybernetic paradigm. A systemic approach to the IoT automatically endows it with another efficiency perspective.

4.5 IoT Availability

In terms of quality/access, many factors are imperative [27]: period, place, delivery of service, network, entity, and device. Figure 9 describes the criteria functionality as well as the challenges faced by those specifications. Augmenting six factors of IoT availability pointed out previously, the environmental facets that are increasingly emerging from the latest development of technology, like green cloud, green computing, including green IoT, must also be considered. This green feature is primarily based on the topic of exhaustible resources and at the same time renewable energy sources.

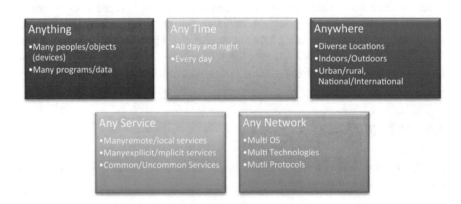

Fig. 9 Availability challenges in cloud computing

4.6 IoT Software

IoT produces vast data volumes that need processing and analysis in real time, which necessitates a large number of activities in data centers, resulting in new security, network capacity, and analytical capability challenges. In addition, the heterogeneity of numerous devices in combination of large data volumes produces challenges for data storage management. These challenges related to IoT data include the nine Vs of big data: value, volatile, validity, veracity, visualization, variability, velocity, variety, and volume (Fig. 10).

4.7 IoT Security

The IoT presupposes the presence of several connected devices, resulting in several access spots and consequently enhanced safety hazards. However, multiple software standards, middleware convergence, application programming interfaces (APIs), machine-to-machine connectivity, and so on unavoidably lead to elevated safety threats and intricate situation [28].

IoT security is equally a concern and a peak priority. We have to guarantee device safety and service security to customers. IoT security analysis has several dimensions incorporating operational security, physical security, IT security, and information security [29]←-[17]. Figure 11 exemplifies few challenges to IoT security. The primary purpose of information security is to protect the secrecy, fairness, and quality of data. Another important security factor in the IoT sense is protection for nonrepudiation functionality of data. This topic remains critical, beyond the four dimensions in IoT data security analysis. In IoT architectures and frameworks, apprehension for quality assurance and standardization development is necessitated due to security concerns. Safety often renders the biggest hurdle along the path of the government-run endeavor for expansion of IoT facilities.

Fig. 10 The addiction for nine Vs of data storage challenges in IoT

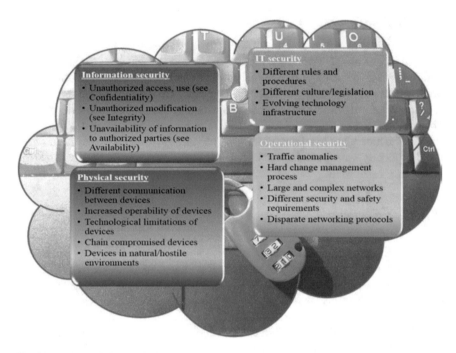

Fig. 11 Security challenges in IoT

The tenaciously diverse disposition of IoT sections leads to the various challenges in security of IoT. Contingent upon these facts, IoT might move each and every one's flaws of the virtual world in the physical world. Bearing in mind major challenges and challenges faced by the connected environment, experts are continuously searching for resolutions.

5 Incorporating IoT and Cloud Computing for Addressing Intrinsic Challenges of IoT

The services proffered through cloud computing convey significant advantages by remodeling storage, analytics, resources, computing, network-adjusted systems and services management, and control and coordination. Service providers provide IoT advantages on stages of every cloud category: Infrastructure as a service (IaaS) enables network and device infrastructure management; platform as a service (PaaS) supports application environment and operating system (OS) management; and software as a service (SaaS) will handle anything connected to users, even applications [30]. Cloud computing in addition provides the IoT with the opportunity of controlling resource admittance through IaaS services; data access through PaaS services; or full admittance toward software applications through SaaS services [31]. IoT solutions delivered by SaaS are developed upon PaaS architecture, allowing business processes through IoT tools and apps to be carried out.

The IoT will allow the most of the three models of cloud computing: private cloud, public cloud, and hybrid cloud. For IoT usage, cloud models to be utilized are selected depending upon explicit necessities and safety. Two convergent methods can be adopted by the joint usage of IoT and cloud computing [32]: introducing IoT technology into the Web (cloud-centric IoT) or putting applications hooked on IoT (IoT-centric). Figures 12 and 13 outline a set of methods for combining IoT with cloud computing along with the categories of cloud services and models. The amalgamation of cloud computing and Internet of Things translates into exemplary literature investigation; for instance, Cloud-IoT, Cloud-based IoT, or Cloud of Things.

5.1 Scalability Via Cloud Computing

An imperative feature assimilated by cloud computing is its versatility in responding to users increasing/diminishing requirements. A function like this allows IoT costs to go down, so consumers disburse only in favor of services used by them. Cloud enables multilevel scalability, such as network management, existing devices, data volume and data storage facility, data diversity, and applications-related services (horizontal and vertical flexibility). A major advantage offered by cloud computing is on-demand scalability. While we scrutinize array of advantages accessible by cloud computing, the power of on-demand scalability is often difficult to conceptualize. Nonetheless, organizations benefit from massive advantages when implementing automated scalability appropriately. It is clear that the scalability advantages fall in compliance with their underlying complexity within the IoT sense. For example, on-demand scalability of only certain apps needs scalability capability across the whole cloud environment (e.g., traffic delivery in multiple instances).

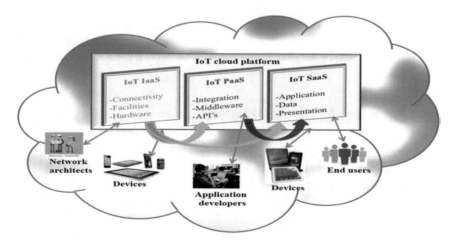

Fig. 12 Cloud-centric IoT

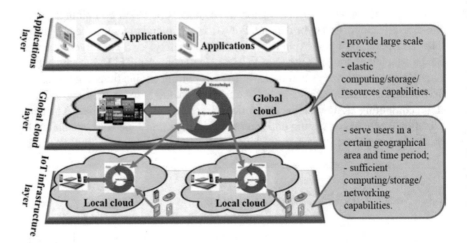

Fig. 13 IoT-centric cloud computing

5.2 Interoperability Through Cloud Computing

IoT comprises an assortment of Internet-connected objects like 2G/3G/fourth generation of wireless communications (4G), Bluetooth, Wi-Fi, NFC, Zigbee, Z-Wave,
and wireless sensor and actuator network (WSAN). In addition, numerous devices
functioning over a long channel create interaction complexity. An alternative is
building hubs, seeing that they can connect through multiple conduits and accumulate signals as of a wide array of devices [33]. Cloud computing extends customary
interfaces and diverse computer portability across the cloud providers [34]. SaaS
providers allow IoT clients to use software directly (over the Internet), from anywhere and devoid of requirement for deploying clandestine servers. It provides hardware autonomy by virtualization, which mitigates device dependence on specific
hardware. PaaS provides interoperable architectures and middleware for services
and data sharing amid diverse devices. Such capabilities for IoT interoperability
have recently ensued by services like container-as-a-service and metal-as-a-service.

5.3 Functionality Via Cloud Infrastructure

Another manner of cloud providing solutions for enhancing system functionality is
by increasing the battery life of devices (e.g., through removing the heavy tasks
assigned to computers) or by developing a scalable architecture [35, 36]. System
efficiency is also enhanced, because cloud computing provides a disruptive-tolerant
technology by improved site-redundant cloud services availability [34]. Cloud computing often utilizes assorted procedures for ensuring data synchronization that
enhances reliability and accuracy in transactions. Good traffic control can increase

network efficiency. Cloud computing provides management procedures that are able to handle unnecessary data transfers and track behaviors for triggering new traffic-sharing instances [37].

5.4 Efficiency Through Cloud Computing

Cloud computing extend various benefits that result in enhanced IoT competence, like multilevel control, enabling availability management, performance, scalability, better energy usage [34], on-demand unlimited processing capabilities, and addressing [38, 39].

5.5 Availability

Data stored in cloud servers are handled uniformly by benchmark APIs [40] and are capable of accessibility and processing from anywhere [41]. Cloud environment provides efficient outcomes for connecting, tracking, and managing anything/object (devices), regardless of time and place, by means of embedded applications and custom gateways [41].

6 Summary and Conclusions

The proliferation of devices with imparting impelling technologies puts the dream of the Internet of Things back together, where tracking and incitement functions regularly blend out from plain sight, and new capabilities are rendered feasible through leveraging rich new data sources. When developing new applications, the creation of the cutting-edge portable platform must depend on the inventiveness of the clients. IoT is a great invention to build that will affect this area by providing new advancing knowledge and the technological tools needed to make innovative applications.

The IoT problem linked to increasing data storage can be addressed by cloud storage, as it extends with flexibility and protection and is capable of customization to IoT network requirements. Cloud computing proffers limitless, inexpensive methods for storing and processing organized as well as unstructured data by implicit storage schemes. Cloud data can be covered by high-level protection implementation [38]. Nevertheless, cloud computing security is even now a significant contest that can despondently proliferate even in the direction of IoT. As pointed out earlier, protection impediments for both IoT and cloud computing are obstacles for growth and widespread acceptance of the two paradigms in more critical fields such as industry.

References

1. S. Shahrestani, H. Cheung, M. Elkhodr, The Internet of Things: Vision & Challenges, in *2013 IEEE Tencon*, (2013), pp. 2018–2222
2. Accenture Strategy, *SMARTer2030: ICT solutions for 21st century challenges*, Belgium Technical Report, 2015 (Global eSustainability Initiative (GeSI), Brussels, 2015)
3. Green Power for Mobile. The Global Telecom Tower ESCO Market. Technical Report 2015
4. G. Fettweis, J. Malmodin, G. Biczok, A. Fehske, The Global footprint of mobile communications: The ecological and economic perspective, 2011. IEEE Communication Magazine
5. IMT Vision-Framework and Overall Objectives of the Future Development of IMT for 2020 and Beyond, 2015, Document Rec. ITU-R M.2083-0
6. M. Albreem, 5G Wireless communication systems: vision and challenges, in *2015 IEEE International Conference on Computer, Communication and Control Technology*, (Malayia, 2015)
7. https://www.ida.gov.sg/~/media/Files/Infocomm%20Landscape/Technology/TechnologyRoadmap/InternetOfThings.pdf
8. D. Norway, E.U. Peter Friess, Belgium Dr. Ovidiu Vermesan SINTEF, Internet of Things: Converging Technologies for Smart Environments and Integrated Ecosystems, in *river publishers' series in communications*, (2013)
9. Norway, Dr. Peter FriessEU, Belgium Dr. Ovidiu Vermesan SINTEF, Internet of Things–From Research and Innovation to Market Deployment, in *River publishers series in communications*, (2014)
10. https://dzone.com/articles/the-internet-of-things-gateways-and-next-generation
11. http://tblocks.com/internet-of-things
12. http://www.reloade.com/blog/2013/12/6characteristics-within-internet-things-iot.php
13. O. Alsaryrah, T.-Y. Chung, C.-Z. Yang, W.-H. Kuo, D.P. Agrawal, I. Mashal, Choices for interaction with things on Internet and underlying issues. Ad Hoc Networks, 68–90 (2015)
14. O. Said, M. Masud, Towards internet of things: survey and future vision. Int. J. Comput. Netw. **5**, 1–17 (2013)
15. T.-J. Lu, F.-Y. Ling, J. Sun, H.-Y. Du, M. Wu, Research on the Architecture of Internet of Things, in *3rd International Conference on Advanced Computer Theory and Engineering (ICACTE '10)*, vol. 5, (Chengdu, China, 2010, August), pp. 484–487
16. S.U. Khan, R. Zaheer, S. Khan, R. Khan, Future internet: the internet of things architecture, possible applications and key challenges, in *10th International Conference on Frontiers of Information Technology (FIT '12)*, (2012, December), pp. 257–260
17. H. Ning, Z. Wang, Future internet of things architecture: like mankind neural system or social organization framework? IEEE Commun. Lett., 461–463 (2011)
18. Y. Bocchi, D. Genoud, A. Jara, The potential of the Internet of Things for defining human behaviours, in *International Conference on Intelligent Networking and Collaborative Systems*, (2014), pp. 581–584
19. S. Chen, S. Xiang, Y. Hu, L. Zheng, Research of architecture and application of Internet of Things for smart grid, in *International Conference on Computer Science and Service System*, (2012), pp. 938–939
20. S. Murugesan, Harnessing green IT: Principles and practices, in *IEEE IT Prof*, (2008), pp. 24–33
21. Gartner, The Impact of the Internet of Things on Data Centers. Gartner report (2014)
22. R. Davis, Big problems with the Internet of Things (2014)
23. Iconectiv, *Overcoming Interoperability Challenges in the Internet of Things* (Telcordia Technologies, 2016)
24. IERC, Internet of Things IoT Semantic Interoperability: Research Challenges, Best Practices, Recommendations and Next Steps (2015)
25. A. Wiles, H. Van der Veer, Achieving technical interoperability, in *The ETSI- approach ETSI*, 3rd edn., (2008)

26. J. Arkko, N. Beheshti, K. Yedavalli J. Kempf, Thoughts on Reliability in the Internet of Things (2011)
27. B. A. Bagula, Internet-of-Things and Big Data: Promises and Challenges for the Developing World, (2016)
28. C. Kocher, The Internet of Things: Challenges and Opportunities, (2014)
29. R. Liwei, IoT Security: Problems, Challenges and Solution, (2015)
30. T. Pasquier, J. Bacon, H. Ko, D. Eyers, J. Singh, Twenty security considerations for cloud-supported internet of things. Internet Things, 269–284 (2016)
31. J. Soldatos, IoT Tutorial: Chapter 4 – Internet of Things in the Clouds, (2016)
32. IoT tutorial, Chapter 4 – Internet of Things in the Clouds, IoT – Internet of Things, (2016)
33. Workflow Studios, Taming the Internet of Things with the Cloud, 2016, March 22, (2016)
34. I. Llorente, Key Challenges in Cloud Computing to Enable Future Internet of Things, (2012)
35. B. Yuxin, M. Yun, Research on the Architecture and Key Technology of Internet of Things (IoT) Applied on Smart Grid, in *Advances in Energy Engineering (ICAEE), International Conference on Advances in Energy Engineering*, (Beijing, 2010), pp. 69–72
36. Z. Bi, L.D. Xu, C. Wang, IoT and Cloud Computing in automation of assembly modeling systems. IEEE Trans Ind Inf **10**(2), 1426–1434
37. R. Adams, E. Bauer, Reliable Cloud Computing – Key Considerations, (2017)
38. S. Mohapatra, P.K. Pattnaik, S.K. Dash, A survey on application of wireless sensor network using Cloud Computing. Int J Comput Sci Eng Technol, 50–55
39. P. Parwekar, From internet of things towards cloud of things, in *Computer and Communication Technology (ICCCT), 2011 2nd international conference on computer and communication technology*, (2011), pp. 329–333
40. S. Kamburugamuve, R.D. Hartman, G.C. Fox, Architecture and measured characteristics of a cloud based Internet of Things, in *Collaboration Technologies and Systems (CTS), 2012 International Conference on Collaboration Technologies and Systems*, (2012), pp. 6–12
41. P. Saluja, N. Sharma, A. Mittal, S.V. Sharma, B.P. Rao, Cloud computing for internet of things & sensing based applications, in *Sensing Technology (ICST), 2012 sixth international conference on sensing technology*, (2012), pp. 374–380

Quantitative Analysis of Industrial IoT System

Rama Kant

1 Introduction

In recent years, formal methods [1], specifically process algebra [2], are widely
used for describing system behaviour and to prove the correctness of IoT's and
modern distributed systems [3–5]. In process algebras [2, 6] framework, there are
two semantic views of the underlying system. First semantic view is related to internal interactions, not seen by external entity, of the system namely called a reduction
semantics [6]. The second semantic view is related to external interactions between
the system and its surrounding environment namely called action semantics [6]. It
interprets the language as a labelled transition system or lts [7, 8]. The labelled
transition system has a very powerful underlying co-inductive proof technique
which is called bisimulation [7, 8]. It induces automatically a bisimulation equivalence between processes. Further, in process algebras framework after defining
these two semantic views, the relation between these two semantic views is to be
found. Routing calculi, DR_π^ω and DR_π [9, 10], are an elaboration of asynchronous
distributed π-calculus [6]. None of these have discussed the quantitative analysis. In
IoT's, which have inherent distributed nature [11, 12], connecting other smart
things, network devices, and servers is the responsibility of the network layer. Its
capabilities are also employed in the transmission and processing of sensor data [11,
12]. The Internet of Things is making distributed computing cool again, according
to [13]. The classic definitions of distributed systems [14, 15] fit industrial IoT systems perfectly. Furthermore, both non-functional and functional criteria occur in
modern distributed networks and industrial IoT systems. Industrial and enterprise
IoT are emerging as attractive solutions for processing IoT applications, thanks to
the rapid development of the IoT. Formal verification approaches, on the other hand,
are necessary to handle and assess critical failures and reachable status in such

R. Kant (✉)
Dr A P J Abdul Kalam Technical University, Lucknow, India

problems due to the guarantee of safety critical conditions without system failures in smart devices [3–5, 15]. The stochastic extension of process algebras [16] and probabilistic extension of process algebras [17, 18] are developed to add quantification to process algebra models. Probability [18] is a key component in modelling and checking quantitative aspects of a system's behaviour. This approach can be extended to a number of sources of uncertainty. We've built GR, a stochastic extension of the [9] routing calculi. The new idea is that the router makes probabilistic assumptions on how to route the message along the communication link between the routers. Because of the instability in real distributed networks, quality of service (QoS) and output metrics (i.e. delay, jitter, latency, etc.) are probabilistic. Because of the probabilistic existence of metrics, the router decision to evaluate the direction from source to destination shifts. So, now we can model more realistic distributed systems with this probabilistic behaviour of router. We can also do performance analysis of distributed networks. In our model, the topology of routers is fixed, i.e. routers connectivity, Γc, is fixed. We assume that router connectivity is bidirectional and each pair of routers is connected via some path. Essentially router connectivity is a connected undirected graph (but not a clique of a graph). So, there may exist more than one path between the same pair of routers. A process, P, resides at a named node n which in turn is connected to router R. The language can calculate the cost and probability of a communication. The cost is calculated by counting the number of hops (routers) a propagating value travels before getting delivered. Here, the cost represents the number of hops (routers) the message has already travelled across the path towards its destination and probability denotes the total probability of the path from source to destination. There are two essential entities of network routing [19, 20]. First is the Routing Protocol which provides each node in a network a consistent view of the topology, and the second is the Routing Algorithm which provides the intelligence to compute path between nodes. Further, it is suggested in [21] that the shortest-hop-path criteria is not sufficient to determine optimal routes. Multiple shortest hop routes may be available, with widely varying levels of reliability. There are many probabilistic routing algorithms developed, for example "An Adaptive Probabilistic Routing Algorithm" [22], PRoPHET [23], "A Probabilistic Emergent Routing Algorithm" [24], and SAMPLE Routing Protocol [25], to achieve qualitative as well as quantitative analysis of real-world distributed networks. In GR, the language for formal verification of such routing protocols has been provided. Earlier, in [26], the formal verification of standard routing algorithms/protocols such as Distance Vector Routing Protocol has been done to prove its correctness. But no routing calculi are available to model probabilistic routing algorithms/protocols with implicit underlying distributed network architecture which is close to real distributed systems. So, the new calculi, GR, proposed in this thesis can be used in formal modelling of the probabilistic routing algorithms/protocols. Note that the other stochastic frameworks in process algebra such as PEPA [27] and PRISM [28] are also available for modelling and quantitative analysis of a distributed system. But, these frameworks have direct connectivity between processes and do not have such implicit formal distributed network architecture. In GR, processes at nodes are connected through routers, i.e. one process P may communicate with another process at the same or different node via routers which is much needed in IoT's and modern distributed system networking requirements. So, new calculus, GR, is developed to model modern distributed systems in a better way by its implicit underlying formal distributed network architecture, also of industrial IoT system [11–13, 15]. To justify the choice of model, GR, an equivalence between GR and its specification PBL [17], will be shown. PBL [17] is most closely related to the motivation for developing a stochastic routing calculi, GR. The different applications of this calculus are given in [17, 29, 30]. A brief introduction of PBL [17] is given in Sect. 2.

2 A Calculus for Probabilistic Wireless Network (PBL)

"A Calculus for Probabilistic Wireless Network (PBL)" [17] is a process calculus used to model high-level wireless systems in which the network topology is represented by a graph. Broadcast connectivity and probabilistic behaviour are aspects of the calculus that are unique to wireless networks. This is designed to model probabilistic protocols such as [25, 31] that are used in wireless networks to enhance network performance. In [17, 29, 31] the numerous implementations of this calculus discussed. Communication can be achieved using a number of channels in this calculus. The correspondence is sent and received in a seamless manner. A system term is essentially a collection of named nodes ranging from $m,n,l,...,$ each of which has some running code (or process). The syntax for this code (or process) is a straightforward instance of a standard process calculus. Process $c?(x).P$ waits to receive some value along channel c; when a value v is received, the process evolves in $\{v/x\}$ Pi; the latter is defined as process P, where all the free occurrences of variable x have been replaced by value v. Process $c!\langle e\rangle.P$ first evaluates a closed expression e to some value v; then this value is sent along channel c, and the process above evolves in P. Here e is some expression from a decidable theory. Process $\tau.P$ performs some internal activity, thus evolving in P. Probabilistic process $P\, p \oplus Q$ behaves as P with probability p, while it behaves as Q with probability $1 - p$. The success process ω is a test process [29]. External nodes can be used to run the test process. The node $m[-]$ is called external node if no code or process has been defined for it. The construct if b the P else Q is the standard if-then-else construct. There are four countable sets in this calculus (Fig. 1).

These sets are **Nodes, Ch, Val,** and **Var** for node names, values, channels, and variables, respectively. The parallel composition operator (.l.) is used to compose different nodes running some code. $eSys$ is a range over all closed well-formed terms in this language, where a well-formed term is viewed as a mapping that assigns node names to the code they are executing. Now we'll look at the reduction semantics of PBL as described in [17]. Before discussing reduction semantics, we provide some mathematical notations for probability distributions. Let S be a set; a function $\Delta\colon S \to [0, 1]$ is called a (probability) sub-distribution over S if $\Sigma_{s\in S}\Delta s \leq 1$. In PBL processes are interpreted as probability distribution of states. Such an

Fig. 1 Syntax of PBL

$$M, N ::=$$ **Nodes**

$n[s]$	Named processes
$M \mid N$	Concurrency
0	Identity

$$P, Q ::=$$ **(Probabilistic)Process**

S	
$P \,_p{\oplus}\, P$	probabilistic choice

$$S, T ::=$$ **states**

$c?(x).P$	receive
$c!\langle e \rangle.P$	broadcast
$S + T$	choice
$if\ b\ then\ S\ else\ T$	branch
$\tau.P$	pre-emption
0	terminate
$\omega.0$	success

(R-STRUCT)

$$\frac{M \equiv M', \Gamma \rhd M' \longrightarrow \Theta, \Theta \equiv \Delta}{\Gamma \rhd M \longrightarrow \Delta}$$

(R-BCAST)

$$\frac{\llbracket e \rrbracket = v \ \ \forall i \in I.\Gamma \vdash m \to n_i \ \ \neg rcv(M, c) \ \ \forall \in nodes(N).\Gamma \vdash m \nrightarrow n}{\begin{array}{l} m\llbracket c!\langle e \rangle.P + Q \rrbracket \prod_{i \in I} |n_i \llbracket (c?(x).P_i) + Q_i \rrbracket \mid M \mid N \longrightarrow \\ \mathbb{P}(m\llbracket P \rrbracket \mid \prod_{i \in I} n_i \llbracket \{^v\!/\!x\}P_i \rrbracket \mid M \mid N) \end{array}}$$

(R-TAU)

$$\frac{}{\Gamma_c \rhd m[\tau.P + Q] \mid M \longrightarrow \mathbb{P}(m[P] \mid M)}$$

Fig. 2 Reduction semantics of PBL

interpretation is encoded by the function P(.) defined as $P(S) = S$ $P((P_1 \,_p{\oplus}\, P_2)) = p.P(P1) + (1 - p.P(P2))$. The reduction semantics of PBL [17] are defined in Fig. 2. Rule (r-bcast) models local broadcast communication.

To justify the model in this chapter, PBL [17] is used as a specification for GR.

3   A Stochastic Routing Calculus

A typical system in, GR, will look like $\langle\, R^\Pi\, \rangle \llbracket n[P] \rrbracket$. A process P is located on a named node n, which is connected to a router R. The process P may communicate with another process on the same node or with another node via a router. Routers serve as a conduit for communication between processes. The routers determine the paths along which the communicated values are propagated with some probability along the router connectivity. Each router maintains a table, called a routing table represented as $\langle\, R^\Pi\, \rangle$, consisting of a mapping for each node name to all adjacent routers, as well as the probability of an output process (i.e. message) occurring along the communication link between routers on the destination node's path. A simple distributed network of routers shown in Fig. 3. There are four routers R_1, R_2, R_3, and R_4. The nodes k, l, m, n, o, r, s, and t are connected to their respective routers R_1, R_2, R_3, and R_4. The notation Adj(R) represents the set of adjacent routers of R in Γc. It is defined as $Adj(R) = \{R'|(R,R') \in \Gamma c\}$. The probability is assigned for output process (i.e. message) along the communication link between routers by the probability distribution Π. We assume that probability distribution, Π, is available for use. We use notation p_{ij} to represent probability of output process along communication link between routers R_i and R_j destination node's paths. The probability distribution Π consists of entries as a triplet (i,j,p_{ij}) where (i,j,p_{ij}) means the output process (i.e. message) hops with probability p_{ij} along the communication link between routers R_i and R_j. In this language, the set of probabilities is provided by probability distribution, Π, and for some router R_i, $\Pi(R_i) = \{(i,j,p_{ij})$, for all R_j such that $R_j \in Adj(R_i)\}$ and $\Sigma p_{ij} = 1$ for all R_j such that $R_j \in Adj(R_i)$. For example, in Fig. 3 for router R_j, $\Pi(R_1) = \{(1, 2, p_{12}), (1, 3, p_{13}), (1, 4, p_{14})\}$ where $p_{12} + p_{13} + p_{14} = 1$. The language can calculate the cost and probability of a communication.

The cost is calculated by counting the number of hops (routers) a propagating value travels before getting delivered. A new syntactic construct called a message is denoted by $[R]_{\boxtimes}^{k,p} \{n,m,v@c\}$ at some router R. Message propagate the value v along channel c from source node n to destination node m. The process at source node n to another process at destination node m can communicate along some channel where n or m may be same. Here, k represents the number of hops (routers) the message has already travelled across the path towards its destination and p denotes the total probability of the path from source to destination. The probability p is calculated by multiplying the probabilities of output process (i.e. message) along the hops a propagating value travels before getting delivered because the probabilities of the output process (i.e. message) along the hops are independent [32]. There are multiple messages reached to destination because router forwards the message to all

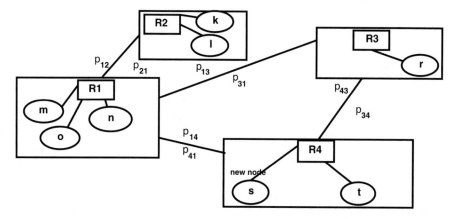

Fig. 3 A simple distributed network with routers

of its adjacent routers. Therefore at the time of delivery of value, duplicate messages are deleted on the basis of cost and probability of received message. After deletion only single message is left which is consumed upon delivery of the value. From the above discussion, it is clear that GR has underlying formal distributed network architecture to model real distributed systems because of probabilistic router. The probabilistic router in GR is now capable to model uncertainty of links in a path from source to destination which is obvious in a real distributed system. Routing table entries are dynamically updated upon creation of new nodes. A method of updating routing tables upon creation of new node is described as in [6]. Only the routing tables that use the new node name in propagating messages or value distribution will be changed. Backward learning update [6] is the name of this technique. The routing table's entries are strongly influenced by router connectivity and probability distribution. The semantics of the systems are therefore described in terms of router connectivity and probability distribution over communication links between routers. Therefore, a configuration, consisting of network of routers, probability distribution, and a system, is defined. The reduction semantics are defined over configuration. Certain conditions on configurations, Γc, $\Pi \triangleright S$, are required for the consistent behaviour of the reduction semantics. A configuration satisfies all these conditions are known as a well-formed configuration. Further, it has proved that well-formedness of a configuration preserved under semantic reductions. To justify our choice of this model, we should establish equivalence of GR with some specification, which has well-established processes algebraic theory, for this model. We show reduction semantic equivalence between "A Calculus for Probabilistic Wireless Network (PBL)" [17], which is briefly explained in Sect. 2, and its implementation, GR. For this purpose, we define a function which maps a system described in new calculus, GR, to a system in "A Calculus for Probabilistic Wireless Network (PBL)" [17]. This function is used to define an invariant property about the system in the new calculus, GR and "A Calculus for Probabilistic Wireless Network (PBL)" [17]. Finally we prove that this invariant is maintained under the reductions of systems described in new calculus, GR and "A Calculus for Probabilistic Wireless Network (PBL)" [17].

4   Syntax

$v, v_1, v_2, u, u_1, u_2, \ldots$ is used to describe either a name or a variable or a simple value. Tuples as values are not used for simplicity. So, here $v, u, \ldots$ are simple values, for example, integers, boolean, etc. Meta variables $a, b, c, \ldots$ range over sets of channel names CN. Variables $m, n, \ldots$ range over set of node names NN. R, R_1, $R_2, \ldots$ range over set of router names RN. To represent the cost of communication, variables $k, l, \ldots$ are used which range over the set of integers. Probability of message hops along communication link between routers are represented through variables $p, p', \ldots$, $p_1, p_2, \ldots$ which range over [0, 1]. It is assumed that sets of channel names, node names, and router names are disjoint from each other. There are three syntactic categories in this language. These are systems, nodes, and processes. The syntax of this calculus is given in Fig. 4. The term system, $\langle R^{\Pi} \rangle \llbracket M \rrbracket$, represents the probabilistic router. The superscript Π represents the probability distribution. This probability distribution is used to assign the probability to output process along communication links between routers. It exhibits the probabilistic nature of the router. The topology of the routers is fixed. The node M is located at router R. Concurrency, S|T, shows the parallel composition of systems S and T. Message is denoted by $\left[\mathbf{R} \right]_{\boxtimes}^{k,p} \{ \boldsymbol{n,m,v} @ \boldsymbol{c} \}$. Message propagates the value v along the channel c from source node n to the destination node m. The process at source node n to another process at destination node m can communicate along some channel. Here, k represents the number of hops (routers) the message has already travelled across the path towards its destination. Here, p denotes the total probability of communication links between the routers the message has already travelled across the path towards its destination. (*new d*) S is the scoping mechanism for names at system level. A process can declare a private channel or a new node name at some router. In this language, it is assumed that router names are not dynamically created. So, d cannot be a router name. The simplest combinator, called the identity function, does nothing but return its argument. The identity for system terms is denoted by ε. In node syntactic category, $n[P]$ represents the named processes. The process term P resides at node name n. The concurrency for nodes terms is denoted by M|N at any router. (*new d*) M is the scoping mechanism of a name at node. A process at a node

Fig. 4 Syntax

S, T ::=		**Systems**
	$\langle R^{\Pi} \rangle \llbracket M \rrbracket$	Probabilistic Router
	$S \mid T$	Concurrency
	$[R]_{\boxtimes}^{k,p} \{n, m, v @ c\}$	Messages
	$(new\ d)\ S$	New Name
	ε	Identity
M, N ::=		**Nodes**
	$n[P]$	Named Processes
	$M \mid N$	Concurrency
	$(new\ d)\ M$	New name
	0	Identity
P, Q ::=		**Process Terms**
	$c?(x)\ P$	Input
	$m!\langle v @ c \rangle$	Output
	$if\ v = u\ then\ P\ else\ Q$	Matching
	$(new\ b)\ P$	Channel name creation
	$newnode\ m\ with\ P\ in\ Q$	New Node Creation
	$P \mid Q$	Concurrency
	$m!\langle v @ c \rangle_{p} \oplus m!\langle v @ c \rangle$	Probabilistic Choice
	$* P$	Repetition
	$stop$	Identity
	ω	Success

may declare a private channel. However, it is possible to export a node name or channel name to another process located at a different node through a value in the message. 0 is the identity for node terms. In process terms, $c?(x)$P is an input process. It receives a value v along the channel c, and then executes P into which v has been substituted, $P\{v/x\}$. Another process is output process, $m!\langle v@c\rangle$. Output process sends a value v to the destination node m on channel c. At destination node m, the value will eventually be delivered to an input process term on channel c. The test for identity of simple values is done by matching construct if $v = u$ then P *else* Q. (*new b*) P is scoping mechanism for names at process level. A process may declare a private channel to communicate within a node. However, it is possible to export a channel name to another process at a different node through a value in message. *newnode m* with P in Q creates a new node at process level. A new node m is created at some router and process P is launched in it in parallel with process Q. The success process ω is a test process [1]. External nodes can be used to run the test process. The node $m[-]$ is called external node if no code or process has been defined for it; for example, $\langle \mathbf{R}^{\Pi}\rangle[\![n[T]|m[c?(x)\ \omega]]\!]$ where $c?(x)\omega$ is a test process. The probabilistic choice between output processes is represented by $m!\langle v@c\rangle\ _p\oplus\ m!\langle v@c\rangle$ i.e. either output process $m!\langle v@c\rangle$ hops with probability p or output process $m!\langle v@c\rangle$ hops with probability $1 - p$. So this means that an output process must hop with total probability 1 because the probability of the second output process is dependent on the first output process. Therefore by rule of probability, total probability is $p + 1 - p = 1$ [31].

Notation 1 In a system $\langle \mathbf{R}^{\Pi}\rangle[\![m[c?(x)P]\!]$, if variable x occurs in the sub-term P, then all occurrences of x in the sub-term P are said to be bound. All occurrences of variables which are not *bound* in a term are said to be *free*.

In Sect. 5, a formal relation between the systems, intuitively to represent the systems as same computational entities, is described, called structural equivalence.

5   Structural Equivalence

Structural equivalence is defined as a formal relation between the systems. Intuitively, structural equivalence is used to represent the systems as the same computational entities. This is defined in a conventional way [2, 6, 9] and the relation is denoted by $\equiv$ in GR. The concept of structural equivalence is that two terms are said to be structurally equivalent if they are same computational entity. In Fig. 5, structure equivalence rules are given that are standard and applicable to all syntactic categories (Figs. 6, 7 and 8).

There are three different syntactic categories for system, nodes, and processes. Therefore, additional axioms for scope extrusion of names from one syntactic category to another are needed. Now, in Sect. 6, the reduction semantics for GR is given. Reduction semantics describe how processes may evolve. The reduction semantics describe the allowable computations from individual processes.

Fig. 5 Structural
equivalence (standard)

(ST-EXTR)	$(new\ d)(P \mid Q)$	$\equiv P \mid (new\ d)Q,\ if\ d \notin fn(P)$
(ST-COM)	$P \mid Q$	$\equiv Q \mid P$
(ST-ASSOC)	$(P \mid Q) \mid R$	$\equiv P \mid (Q \mid R)$
(ST-ID)	$P \mid id$	$\equiv P$
(ST-FLIP)	$(new\ c)(new\ d)P$	$\equiv (new\ d)(new\ c)P$
(ST-NEW)	$(new\ d)P$	$\equiv P,\ if\ d \notin fn(P)$
(ST-NEW $-$ ID)	$(new\ d)id$	$\equiv id$

(SP-STD)	*standard axioms*	
(SP-PROBOUT)	$m!\langle v@c \rangle_p \oplus m!\langle v@c \rangle$	$\equiv m!\langle v@c \rangle_{p'} \oplus m!\langle v@c \rangle,\ p' = 1 - p$
	$m!\langle v@c \rangle_p \oplus m!\langle v@c \rangle$	$\equiv m!\langle v@c \rangle\ if\ p = 1$
(SP-REP)	$*P$	$\equiv P \mid *P$

Fig. 6 Structural equivalence (processes)

(SN-STD)	*standard axioms*		
(SN-STOP)	$m[stop]$	$\equiv$	0
(SN-INH)		$\dfrac{P \equiv Q}{m[P] \equiv m[Q]}$	
(SN-EXTR)	$m[(new\ d)P]$	$\equiv$	$(new\ d)m[P],\ if\ d \neq m$
(SN-MERGE)	$m[P] \mid m[Q]$	$\equiv$	$m[P \mid Q]$

Fig. 7 Structural equivalence (nodes)

Fig. 8 Structural
equivalence (systems)

(SS-STD)	*standard axioms*
(SS-INH)	$\dfrac{N \equiv S}{\begin{array}{c}\langle R^{\Pi} \rangle [\![N]\!] \equiv \langle R^{\Pi} \rangle [\![S]\!] \\ \langle R^{\Pi_1} \rangle [\![N]\!] \equiv \langle R^{\Pi_2} \rangle [\![S]\!]\end{array}}$
(SS-EXTR)	$(new\ d)\langle R^{\Pi} \rangle [\![N]\!] \equiv \langle R^{\Pi} \rangle [\![(new\ d)N]\!]\ ,if\ d \neq R$

6 Reduction Semantics

In this calculi, each router maintains routing table, $\langle R^{\Pi} \rangle$, and determines the paths of communication along which the value to be communicated propagates with some probability. The routing table entries are strongly influenced by router connectivity, probability distribution, and the system itself. As a consequence, the systems' reduction semantics are defined in terms of router connectivity and the probability distribution over communication links between routers. It's a binary relationship between structures known as configurations. The network of routers and probability distribution are used to describe a configuration. The network of routers Γc is basically a binary relation between the router names. For example, in Fig. 3, a network of routers Γc may be defined as set of pairs {(R_1,R_1), (R_1,R_2), (R_1,R_3),(R_1,R_4), (R_2,R_2), (R_3,R_1), (R_3,R_4), (R_3,R_3), (R_4,R_1), (R_4,R_3), (R_4,R_4)}. For the simplicity of the model, the reflexive pairs can be removed from Γc because a router is always connected to itself. This does not affect our basic objective of showing the quality of service and performance of the network in a particular communication between the processes because in this calculus the quality of service and performance of the network is demonstrated by the number of hop (router) crossings a message does before delivering a value along a chosen path and the probability of the chosen path. Thus $\Gamma c = \{(R_1,R_2), (R_1,R_3),(R_1,R_4), (R_3,R_1), (R_3,R_4), (R_4,R_1), (R_4,R_3)\}$ with the property that network of routers Γc is undirected. Further we assume that Γc is connected. This property means that all pairs of routers in the network are connected by at least one path. The configuration $\Gamma c,\Pi \rhd S$, where Γc is the network of routers connectivity and Π is the probability distribution. It is defined below:

Configuration 1 A configuration consists of a triplet $(\Gamma c,\Pi,S)$ where $\Gamma c \subseteq RN \times RN$, the network of routers connectivity, Π is the probability distribution for output process along communication links between routers and is the set of router names in any system S. The undirected graph $(RN, \Gamma c)$ is connected.

Now we use following notation for path between routers:

Notation 2 In a router connectivity Γc, for routers $R_i, R_{i+1},\ ..,\ R_{k-1}, R_k, R_i \leadsto R_{i+1} \leadsto\ ... \leadsto R_{k-1} \leadsto R_k$ means $\{R_i, R_{i+1},\ ..,\ R_{k-1}, R_k\} \in \Gamma c$. We shall use the notation **path**(R_i, R_k) to represent a path of routers, $R_i \leadsto\ ... \leadsto R_k$ between pair of routers R_i and R_k.

In this calculus, it must be ensured that at the time of delivery of the value carried by the message at router of destination node, only one message is present. Using the following notation for this:

Notation 3 S' notion S means if $S \equiv S_1|S_2|...|S_n$ then $S' \neq S_i'$ such that $S_i \equiv S_i'$ for all i = 1,2,3,...,n.

The reduction semantics of the systems are defined in terms of a binary relation $- \rightarrow^{k,p}$ between the configurations where the cost of reduction is k and the probability of reduction is p. We represent the structural equivalence between configurations by stripping off the router connectivity. For the purpose of determining the path between the communicating processes routing table, $\langle R^\Pi \rangle$, is available at each router, R. Formally, the routing table, $\langle R^\Pi \rangle$, is defined as follows:

Definition 1 In a configuration $\Gamma c,\Pi \rhd \langle R^\Pi \rangle [\![N]\!] |S$, the named router table $\langle R^\Pi \rangle$ is defined as a function $\langle R^\Pi \rangle: NN \rightarrow 2^{RN}$ where a set of node names, NN and RN, is a set of router names.

The function $\langle R^\Pi \rangle$ can be many-to-many. Several node names may have the same adjacent router on the path towards the destination and each destination node will have many choices of adjacent routers. If a node name m belongs to the domain of the routing table $\langle R^\Pi \rangle$, then the notation $\langle R^\Pi \rangle (m) \downarrow, m \in NN$ shall be used. Similarly if the node name m does not belong to the domain of the routing table $\langle R^\Pi \rangle$, then we use the notation $\langle R^\Pi \rangle (m) \uparrow, m \in NN$. In this process algebra, a judgement $\Gamma c,\Pi \rhd S- \rightarrow^{k,p} \Gamma c, \Pi \rhd S'$ would intuitively mean that a system S reduces to S' where the cost of the reduction is k and probability of the reduction is p. Reduction rules are given in Figs. 9 and 10.

The reduction rule (r-out) generates a propagation message $[R]_\boxtimes^{0,1} \{n,m,v@c\}$ in parallel with the system $\langle R^\Pi \rangle [\![n[P]|N]\!]$. The message with subscripts 0 and 1 indicates that it has been generated at the router R. The message is not hopped at any other router yet. The message will propagate the value v to the destination node m. The reduction rules (r-bcom - i) and (r-bcom - ii) are used to propagate the message from one router to the other towards the destination node. If the message is carrying node name and is unknown to intermediate nodes, then routing tables of intermediate routers must be updated about this node information. This is called backward learning updates, Example 1 shall demonstrate this. The delivery of the message at the destination node is done by reduction rule (r-in). The delivery is done when only single message is left at the destination router (Fig. 11).

Example 1 As shown in Fig. 12, we have the following probabilities assigned for output process (i.e. messages) along communication links between routers provided by probability distribution $\Pi = \{(1, 2, 0.50), (1, 3, 0.30), (1, 5, 0.20), (2, 1, 1), (3, 1, 0.60), (3, 4, 0.40),; (4, 3, 0.30), (4, 5, 0.70), (5, 1, 0.7), (5, 4, 0.30)\}$. The systems S_1, S_2, S_3, S_4, and S_5 are below:

$$S_1 \equiv \langle\ R_1^{\Pi}\ \rangle [\![m[c?\ (x) t! \langle\ x@d\rangle] | n[N] | o[O]\ ]\!],$$
$$S_2 \equiv \langle\ R_2^{\Pi}\ \rangle [\![N_2\ ]\!],$$
$$S_3 \equiv \langle\ R_3^{\Pi}\ \rangle [\![N_3\ ]\!],$$
$$S_4 \equiv \langle\ R_4^{\Pi}\ \rangle [\![\ s[new\ node\ q\ with\ P\ in\ Q]\ ]\!],$$
$$S_5 \equiv \langle\ R_5^{\Pi}\ \rangle [\![t[d?(x)Q]\ ]\!],\ \text{where}\ P \equiv m! \langle\ q@c\rangle | Q'.$$

A new node q is created at node s. This new node q is sent to process at node m. So first by an application of (r-newnode - creation), the following configuration is received.

$\Gamma c, \Pi \triangleright S_1 | S_2 | S_3 | \langle\ R_4^{\Pi}\ \rangle [\![\ s[new\ node\ q\ with\ P\ in\ Q]\ ]\!], | S_5$ reduces to $\Gamma c, \Pi \triangleright S_1 | S_2 | S_3 | (new\ q)(\langle\ R_4^{\Pi}\ \{q \to R4\}\ \rangle [\![\ q[P]\ | s[Q]\ ]\!]) | S_5$. By the use of (r-struct), we get $\Gamma c, \Pi \triangleright S_1 | S_2 | S_3 | (new\ q)(\langle\ R_4^{\Pi}\ \{q \to R4\}\ \rangle [\![\ q[m! \langle\ q@c\rangle | Q']\ | s[Q]\ ]\!]) | S_5$. Further, by using reduction rule (r-out), the configuration $\Gamma c, \Pi \triangleright S_1 | S_2 | S_3 | (new\ q)(\langle\ R_4^{\Pi}\ \{q \to R4\}\ \rangle [\![\ q[m! \langle\ q@c\rangle | Q']\ | s[Q]\ ]\!]) | S_5$ reduces to $\Gamma c, \Pi \triangleright S_1 | S_2 | S_3 | (new\ q)(\langle\ R_4^{\Pi}\ \{q \to R4\}\ \rangle [\![\ q[Q']\ | s[Q]\ ]\!]) | [R4]_{\boxtimes}^{0,1}\ \{q, m, q@c\}\ | S_5$. Now, $\langle\ R_4^{\Pi}\ (m)\rangle = \{R_3, R_5\}$. q is a new node and is not known to R_4, R_5, i.e. $\langle\ R_3^{\Pi}\ (m)\rangle\ (q)\ \uparrow, q \in NN$ and $\langle\ R_5^{\Pi}\ (m)\rangle\ (q)\ \uparrow, q \in NN$. By reduction rules (r-bcom - i) and (r-struct), the configuration,

$\Gamma c, \Pi \triangleright S_1 | S_2 | S_3 | (new\ q)(\langle\ R_4^{\Pi}\ \{q \to R4\}\ \rangle [\![\ q[Q']\ | s[Q]\ ]\!]) | [R4]_{\boxtimes}^{0,1}\ \{q, m, q@c\}\ | S_5$ reduces to $\Gamma c, \Pi \triangleright S_1 | S_2 | \langle\ R_3^{\Pi}\ \{q \to R3\}\ \rangle [\![N_3\ ]\!]\ | (new\ q)(\langle\ R_4^{\Pi}\ \{q \to R4\}\ \rangle [\![\ q[Q']\ | s[Q]\ ]\!]) | [R5]_{\boxtimes}^{1,0.7}\ \{q, m, q@c\}\ |\ [R3]_{\boxtimes}^{1,0.3}\ \{q, m, q@c\}\ | \langle\ R_5^{\Pi}\ \{q \to R5\}\ \rangle [\![t[d?(x)Q]\ ]\!].$

Now, $\langle\ R_3^{\Pi}\ (m)\rangle = \{R_1\}$ and $\langle\ R_5^{\Pi}\ (m)\rangle = \{R_1\}$ where $\langle\ R_1^{\Pi}\ (m)\rangle\ (q)\ \uparrow, q \in NN$. By the use of (r-bcom - i) and (r-struct), the configuration $\Gamma c, \Pi \triangleright \langle\ R_1^{\Pi}\ \rangle [\![m[c?\ (x) t! \langle\ x@d\rangle]\ | n[N]\ | o[O]\ ]\!]\ | S_2 | \langle\ R_3^{\Pi}\ \{q \to R3\}\ \rangle [\![N_3\ ]\!]\ | (new\ q)(\langle\ R_4^{\Pi}\ \{q \to R4\}\ \rangle [\![\ q[Q']\ | s[Q]\ ]\!]) | [R5]_{\boxtimes}^{1,0.7}\ \{q, m, q@c\}\ |\ [R3]_{\boxtimes}^{1,0.3}\ \{q, m, q@c\ | \langle\ R_5^{\Pi}\ \{q \to R5\}\ \rangle [\![t[d?(x) Q]\!]$ reduces to $\Gamma c, \Pi \triangleright \langle\ R_1^{\Pi}\ \{q \to R1\}\ \rangle [\![m[c?\ (x) t! \langle\ x@d\rangle]\ | n[N]\ | o[O]\ ]\!]\ | S_2 | \langle\ R_3^{\Pi}\ \{q \to R3\}\ \rangle [\![N_3\ ]\!]\ | (new\ q)(\langle\ R_4^{\Pi}\ \{q \to R4\}\ \rangle [\![q[Q'] | s[Q]]\!]) | [R1]_{\boxtimes}^{2,0.49}\ \{q, m, q@c\}\ |\ [R1]_{\boxtimes}^{2,0.18}\ \{q, m, q@c\ | \langle\ R_5^{\Pi}\ \{q \to R5\}\ \rangle\ [\![t[d?(x) Q]]\!]$. Now messages have reached the destination router, i.e. $\langle\ R_1^{\Pi}\ (m)\rangle = \{R_1\}$. So, the duplicate messages are deleted by (r-msg - del), and the following is received: $\Gamma c, \Pi \triangleright \langle\ R_1^{\Pi}\ \{q \to R1\}\ \rangle [\![m[c?\ (x) t! \langle\ x@d\rangle]\ | n[N]\ | o[O]\ ]\!]\ | S_2 | \langle\ R_3^{\Pi}\ \{q \to R3\}\ \rangle [\![N_3\ ]\!]\ | (new\ q)(\langle\ R_4^{\Pi}\ \{q \to R4\}\ \rangle [\![$

(R-OUT)

$$\Gamma_c, \Pi \; \triangleright \; \langle R^{\Pi} \rangle [\![n[m!\langle v@c \rangle \mid P] \mid N]\!] \longrightarrow \Gamma_c, \Pi \; \triangleright \; [R]_{\boxtimes}^{0,1} \{n, m, v@c\} \mid \langle R^{\Pi} \rangle [\![n[P] \mid N]\!]$$

(R-BCOM−I)

$$\{(R_1, R_j)\} \in \Gamma_c$$
$$p_{1j} > 0, p_{1j} \in \Pi$$
$$\langle R_1^{\Pi} \rangle (m) = \{R_j\}$$
$$\langle R_j^{\Pi} \rangle (v) \uparrow, v \in \mathcal{NN}$$
$$\rule{8cm}{0.4pt} \text{ for all } j=2,3,..$$
$$\Gamma_c, \Pi \; \triangleright \; [R_1]_{\boxtimes}^{k,p} \{n, m, v@c\} \mid \langle R_j^{\Pi} \rangle [\![N]\!] \mid S \longrightarrow$$
$$\Gamma_c, \Pi \; \triangleright \; [R_j]_{\boxtimes}^{k+1, p*p_{1j}} \{n, m, v@c\} \mid \langle R_j^{\Pi} \{v \to R_1\} \rangle [\![N]\!] \mid S$$

(R-BCOM−II)

$$\{(R_1, R_j)\} \in \Gamma_c$$
$$p_{1j} > 0, p_{1j} \in \Pi$$
$$\langle R_1^{\Pi} \rangle (m) = \{R_j\}$$
$$\langle R_j^{\Pi} \rangle (v) \downarrow, v \in \mathcal{NN}$$
$$\rule{8cm}{0.4pt} \text{ for all } j=2,3,..$$
$$\Gamma_c, \Pi \; \triangleright \; [R_1]_{\boxtimes}^{k,p} \{n, m, v@c\} \mid \langle R_j^{\Pi} \rangle [\![N]\!] \mid S \longrightarrow$$
$$\Gamma_c, \Pi \; \triangleright \; [R_j]_{\boxtimes}^{k+1, p*p_{1j}} \{n, m, v@c\} \mid \langle R_j^{\Pi} \rangle [\![N]\!] \mid S$$

(R-MSG−DEL)

$$\langle R^{\Pi} \rangle (m) = \{R\}$$
$$\rule{9cm}{0.4pt}, \text{if } k < k'$$
$$\Gamma_c, \Pi \; \triangleright \; [R]_{\boxtimes}^{k,P} \{n, m, v@c\} \mid [R]_{\boxtimes}^{k',p'} \{n, m, v@c\} \mid S \longrightarrow$$
$$\Gamma_c, \Pi \; \triangleright \; [R]_{\boxtimes}^{k,P} \{n, m, v@c\} \mid S$$

$$\langle R^{\Pi} \rangle (m) = \{R\}$$
$$\rule{9cm}{0.4pt}, \text{if } k > k'$$
$$\Gamma_c, \Pi \; \triangleright \; [R]_{\boxtimes}^{k,P} \{n, m, v@c\} \mid [R]_{\boxtimes}^{k',p'} \{n, m, v@c\} \mid S \longrightarrow$$
$$\Gamma_c, \Pi \; \triangleright \; [R]_{\boxtimes}^{k',p'} \{n, m, v@c\} \mid S$$

$$\langle R^{\Pi} \rangle (m) = \{R\}$$
$$\rule{9cm}{0.4pt}, \text{if } k = k' \text{and } p > p'$$
$$\Gamma_c, \Pi \; \triangleright \; [R]_{\boxtimes}^{k,P} \{n, m, v@c\} \mid [R]_{\boxtimes}^{k',p'} \{n, m, v@c\} \mid S \longrightarrow$$
$$\Gamma_c, \Pi \; \triangleright \; [R]_{\boxtimes}^{k,P} \{n, m, v@c\} \mid S$$

$$\langle R^{\Pi} \rangle (m) = \{R\}$$
$$\rule{9cm}{0.4pt}, \text{if } k = k' \text{and } p < p'$$
$$\Gamma_c, \Pi \; \triangleright \; [R]_{\boxtimes}^{k,P} \{n, m, v@c\} \mid [R]_{\boxtimes}^{k',p'} \{n, m, v@c\} \mid S \longrightarrow$$
$$\Gamma_c, \Pi \; \triangleright \; [R]_{\boxtimes}^{k',p'} \{n, m, v@c\} \mid S$$

$$\langle R^{\Pi} \rangle (m) = \{R\}$$
$$\rule{9cm}{0.4pt}, \text{if } k = k' \text{and } p = p'$$
$$\Gamma_c, \Pi \; \triangleright \; [R]_{\boxtimes}^{k,P} \{n, m, v@c\} \mid [R]_{\boxtimes}^{k',p'} \{n, m, v@c\} \mid S \longrightarrow$$
$$\Gamma_c, \Pi \; \triangleright \; [R]_{\boxtimes}^{k,P} \{n, m, v@c\} \mid S$$

(R-IN)

$$\langle R^{\Pi} \rangle (m) = \{R\}, [R]_{\boxtimes}^{l,q} \{n, m, v@c\} \; \textbf{notin} \; S \text{ for any } l \text{ and } q$$
$$\rule{9cm}{0.4pt}$$
$$\Gamma_c, \Pi \; \triangleright \; [R]_{\boxtimes}^{k,P} \{n, m, v@c\} \mid \langle R^{\Pi} \rangle [\![m[c?(x) \, P] \mid N]\!] \mid S \longrightarrow^{k,p}$$
$$\Gamma_c, \Pi \; \triangleright \; \langle R^{\Pi} \rangle [\![m[P\{v/x\}] \mid N]\!] \mid S$$

Fig. 9 Reduction Semantics (Contd...)

(R-NEWNODE−CREATION)
$\Gamma_c, \Pi \rhd \langle R^{\Pi} \rangle [\![n [newnode\ m\ with\ P\ in\ Q]]\!] \longrightarrow$
$\Gamma_c, \Pi \rhd (new\ m)(\langle R^{\Pi} \{ m \rightarrow R \} \rangle [\![m[P] \mid n[Q]]\!])$

(R-MATCH)
$\Gamma_c, \Pi \rhd \langle R^{\Pi} \rangle [\![n[if\ v = u\ then\ P\ else\ Q]]\!] \longrightarrow \Gamma_c, \Pi \rhd \langle R^{\Pi} \rangle [\![n[P]]\!]$

(R-MISMATCH)
$\Gamma_c, \Pi \rhd \langle R^{\Pi} \rangle [\![n[if\ v_1 = v_2\ then\ P\ else\ Q]]\!] \longrightarrow \Gamma_c, \Pi \rhd \langle R^{\Pi} \rangle [\![n[Q]]\!] \quad v_1 \neq v_2$

(R-STRUCT)
$$\frac{S \equiv S', \Gamma_c, \Pi \rhd S' \longrightarrow^{k,p} \Gamma_c, \Pi \rhd R', R' \equiv R}{\Gamma_c, \Pi \rhd S \longrightarrow^{k,p} \Gamma_c, \Pi \rhd R}$$

(R-CNTX)
$$\frac{\Gamma_c, \Pi \rhd S_1 \longrightarrow^{k,p} \Gamma_c, \Pi \rhd S_1'}{\begin{array}{l} \Gamma_c, \Pi \rhd S_1 \mid S_2 \longrightarrow^{k,p} \Gamma_c, \Pi \rhd S_1' \mid S_2 \\ \Gamma_c, \Pi \rhd S_2 \mid S_1 \longrightarrow^{k,p} \Gamma_c, \Pi \rhd S_2 \mid S_1' \\ \Gamma_c, \Pi \rhd (new\ d)S_1 \longrightarrow^{k,p} \Gamma_c, \Pi \rhd (new\ d)S_1' \end{array}}$$

Fig. 10 (Contd....) reduction semantics

Fig. 11 Adapted
(r-bcom - ii)

$$\frac{\{(R_1, R_j)\} \in \Gamma_c \\ p_{1j} > 0, p_{1j} \in \Pi \\ \langle R_1^{\Pi} \rangle (m) = \{R_j\} \\ \Gamma_c, \Pi \rhd [R_1]_{\boxtimes}^{k,p} \{n, m, v@c\} \mid \langle R_j^{\Pi} \rangle [\![N]\!] \mid S \longrightarrow}{\Gamma_c, \Pi \rhd [R_j]_{\boxtimes}^{k+1, p*p_{1j}} \{n, m, v@c\} \mid \langle R_j^{\Pi} \rangle [\![N]\!] \mid S} \text{for } j=2,3,\ldots$$

$q[Q'] \mid s[Q] \;]\!])\mid [R1]_{\boxtimes}^{2,0.49} \{q,m,q@c\} \mid \langle R_5^{\Pi} \{q \rightarrow R5 \rangle [\![t[d?(x)Q]]\!]$. By using (r-in), the value is delivered to input process, $c?(x)$; at node s, we get the following reduction: $\Gamma c, \Pi \rhd \langle R_1^{\Pi} \{q \rightarrow R1\} \rangle [\![m[c?\ (x)t!\langle x@d \rangle] \; ln[N] \; lo[O]]\!] \mid S_2 \mid \langle R_3^{\Pi} \{q \rightarrow R3\} \rangle [\![N_3]\!] \mid (new\ q)(\langle R_4^{\Pi} \{q \rightarrow R4\} \rangle [\![q[Q'] \; \mid s[Q] \;]\!])\mid [R1]_{\boxtimes}^{2,0.49} \{q,m,q@c\} \mid \langle R_5^{\Pi} \{q \rightarrow R5 \rangle [\![t[d?(x)Q]]\!] \rightarrow^{2,0.49} \Gamma c, \Pi \rhd \langle R_1^{\Pi} \{q \rightarrow R1\} \rangle [\![m[t!\langle q@d \rangle] \; ln[N] \; lo[O]]\!] \mid S_2 \mid \langle R_3^{\Pi} \{q \rightarrow R3\} \rangle [\![N_3]\!] \mid (new\ q)(\langle R_4^{\Pi} \{q \rightarrow R4\} \rangle [\![q[Q'] \; \mid s[Q] \;]\!])\mid \langle R_5^{\Pi} \{q \rightarrow R5 \rangle [\![t[d?(x)Q]]\!]$.

Similarly, this node name q further sends to process at node t by output process t!⟨q@d⟩ at node m by using (r-struct), (r-out), and (r-bcom-ii).

In Example 1, the use of reduction semantics to send simple values as well as new node names through message during communication between processes at nodes has been discussed. The backward learning updates for updating routing table entries on new node creation has been also discussed. The duplicate message deletion at the destination router is also discussed. Reductions are preserved up to structural equivalence. This is formally stated and proved in the following theorem:

Theorem 1 If $S_1 \equiv S_2$ and $\Gamma c, \Pi \rhd S_1 \rightarrow \Gamma c, \Pi \rhd S_1'$, then there exist S_2' such that $\Gamma c, \Pi \rhd S_2 \rightarrow \Gamma c, \Pi \rhd S_2'$ s.t. $S_1' \equiv S_2'$.

In this section we have discussed the reduction semantics with the help of two examples. We also proved that reduction semantics is preserved under structural equivalence. Next, how will we ensure that after applying a reduction semantics on

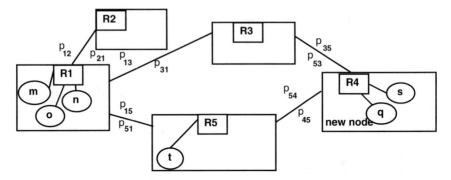

Fig. 12 A simple distributed network with routers (new node creation)

configurations in GR, there is no inconsistency in residual configuration after a reduction? In Sect. 7, we discuss about these conditions called well-formed conditions for configurations in GR.

7   Well-Formed Configurations

In our calculi, how can we say that reduction semantics is reasonable? How will we ensure that there is no inconsistency in resulting configuration? We can say if a configuration is coherent before we apply a reduction and it remains coherent after the reduction as well. We can list the properties which will ensure coherence, together gets us the notion of well-formed configurations, which are summarized in Definition 2.

Definition 2 (Well-formed configuration).

1. $\Gamma c, \Pi \triangleright S$ is a well-formed system.
2. $\Gamma c, \Pi \triangleright (new\ n)S$ is well formed if $\Gamma c, \Pi \triangleright S$ is a well-formed system.
3. $\Gamma c, \Pi \triangleright \langle R^{\Pi} \rangle [\![N]\!] | S$ is well formed if.

   (a) $\Gamma c, \Pi \triangleright S$ is well formed.
   (b) $\langle R^{\Pi} \rangle$ does not occur in S (unique router name).
   (c) $\forall r \in fn(N)$ such that $r \in NN$ where NN is the set of node names, if $\langle R^{\Pi} \rangle$ $(r) = \{R\}$ then $\forall \langle R^{\Pi} \rangle \in S, \langle R_1^{\Pi} \rangle(r) \neq \{R_1\}$ (unique node name).
   (d) $\forall r \in fn(N)$ such that $r \in NN, \langle R \rangle(r) \downarrow$.

4. $\Gamma c, \Pi \triangleright [R]_{\boxtimes}^{k,p} \{n,m,v @ c\}$ |S well formed if.

   (a) $\Gamma c, \Pi \triangleright S$ is well formed and $S \equiv \langle R^{\Pi} \rangle [\![N]\!] | S'$ for some S'.
   (b) $\forall r \in \{m\} \bigcup fn(v)$ such that $r \in NN, \langle R \rangle(r) \downarrow$.
   (c) $\Gamma c, \Pi \triangleright S$ is well formed if S is of the form $S_1 | S_2$ where $S_1 \equiv [R]_{\boxtimes}^{k1,p1} \{n,m,v @ c\}$ | $[R]_{\boxtimes}^{k2,p2} \{n,m,v @ c\}$ | $[R]_{\boxtimes}^{k3,p3} \{n,m,v @ c\}$ |... for $R_1, R_2, R_3, ...$ for some S_2.

(d) There exists path $\textbf{path}(R_1,R) = R_1 \rightsquigarrow R_2 \rightsquigarrow \ldots \rightsquigarrow R$ for some $R_1, R_2, \ldots$ such that $\langle\ R_1^{\Pi}\ \rangle(n) = \{R_1\}$ and $\langle\ R_1^{\Pi}\ \rangle(m) = \{R_2\}$ ... where $k = |\textbf{path}(R_1,R)|- 1$.

5. A configuration $\Gamma c, \Pi \vartriangleright S$ is well formed if $\langle\ R_1^{\Pi}\ \rangle, \langle\ R_j^{\Pi}\ \rangle$ occur in S and $\langle\ R_1^{\Pi}\ \rangle(m) = \{R_j\}$, and then $\langle R_j \rangle(m) \downarrow$ for all $j = 2, 3, 4....$

6. In any well-formed configuration $\Gamma c, \Pi \vartriangleright S$ for every pair of node n and m such that $\langle\ R_n^{\Pi}\ \rangle(n) = \{R_n\}$ and $\langle\ R_m^{\Pi}\ \rangle(m) = \{R_m\}$ where R_n, R_m are in S, there exists a path $R_n \rightsquigarrow R_m$ such that

$$\langle\ R_n^{\Pi}\ \rangle(m) = \{R_1\}, \langle\ R_1^{\Pi}\ \rangle(m) = \{R_2\},..., \langle\ R_m^{\Pi}\ \rangle(m) = \{R_m\}.$$

We have the following lemma to show the well-formedness of configurations is preserved up to structural equivalence of its systems:

Lemma 1 Suppose $S \equiv T$, then $\Gamma c, \Pi \vartriangleright S$ is well formed if and only if $\Gamma c, \Pi \vartriangleright T$ is well,formed.

Additionally, well-formedness of configurations is also preserved under reduction semantics as given in the following theorem:

Theorem 2 If $\Gamma c, \Pi \vartriangleright S$ is a well-formed configuration and $\Gamma c, \Pi \vartriangleright S\text{-} \rightarrow {}^{k,p} \Gamma c, \Pi \vartriangleright S'$, then $\Gamma c, \Pi \vartriangleright S'$ is also well formed.

We'll now try to find an equivalence between GR and its specification, "A Calculus for Probabilistic Wireless Network (PBL)" [17] in Sect. 8.

8   Equivalence of GR with Its Specification

To justify our choice of this model, we should establish equivalence of GR with some specification, which has well-established processes algebraic theory, for this model. Therefore, in this section we show reduction semantic equivalence between "A Calculus for Probabilistic Wireless Network (PBL)" [17] and its implementation, GR. We define a function for new calculus which maps a system described in new calculus, GR, to a system in "A Calculus for Probabilistic Wireless Network (PBL)" [17]. This function is used to define an invariant property about the system in the new calculus, GR and "A Calculus for Probabilistic Wireless Network (PBL)" [17]. Finally we prove that this invariant is maintained under the reductions of systems described in new calculus, GR and "A Calculus for Probabilistic Wireless Network (PBL)" [17]. We will show that both languages, GR and PBL, are reduction equivalent after abstracting away the details of routers and paths from GR. Firstly, we define a function, F, in Definition 3 over GR system to map to a system in PBL for the purpose of abstraction of routers.

Definition 3 We define a function $F,: \text{GRY} \rightarrow \text{PBLY}$, where GRY and PBLY are sets of GR system terms and PBL system terms, respectively, as follows:

$$\begin{aligned}
\mathcal{F}(\varepsilon) &= 0 \\
\mathcal{F}(\langle R^{\Pi} \rangle \llbracket M \rrbracket) &= M \\
\mathcal{F}([R]_{\boxtimes}^{k,p}\{n, m, v@c\}) &= n[m!\langle v@c \rangle_p \oplus m!\langle v@c \rangle] \\
\mathcal{F}((newd)\,S) &= (new\ d)\mathcal{F}(S) \\
\mathcal{F}(S|T) &= \mathcal{F}(S)|\mathcal{F}(T)
\end{aligned}$$

This function, F, can be applied to S in GR and T in PBL up to structural equivalence. The first base rule of function is $F(\varepsilon) = 0$ which maps the identity term, ε, of GR into identity term, 0, of PBL. In the system term of GR, we abstract the routers and map it to node which is the system syntactic category in PBL. This is the second base rule for function. In the third rule, the system term message in GR is mapped to node term in PBL with probabilistic output process in it. This essentially means that in PBL, probabilistic output process term at respective node n is equivalent to corresponding message at respective router R in GR. This is intuitively correct because the GR system term messages are generated by an output process term using the reduction rule (r-out) in Fig. 9. The messages basically output a value to some specified channel located at some node with some probability. Other rules are inductive. To prove this equivalence, we need propositions 1 and 2.

Proposition 1 For any system term S in GR such that $F(S) = T$ and $T \equiv T'$, for some system term T and T' in PBL, implies that there exists some system term S' in GR such that $(S') = T'$ and $S \equiv S'$.

Proposition 2 For any system term S in GR such that $S \equiv S'$ implies $(S) \equiv (S')$.

Further, we give the following theorems, 3 and 4, to prove this equivalence. In Theorem 3, we prove that whenever a well-formed configuration in GR does a reduction, there exists a corresponding PBL [17] system.

Theorem 3 In GR, if a well-formed configuration $\Gamma c, \Pi \triangleright S_1$ does a reduction $\Gamma c, \Pi \triangleright S_1 \to {}^{k,p} \Gamma c, \Pi \triangleright S_2$ for some k and p and $F(S_1) = T_1$ where T_1 is a PBL system term then:

- Either there exists a PBL system T_2 s.t. $T_1 \to T_2$ and $F(S_2) \equiv T_2$.
- Or $F(S_2) \equiv T_1$.

Theorem 4 If a PBL system T_1 does a reduction $T_1 \to T_2$ and $F(S_1) = T_1'$ such that $T_1' \equiv T_1$ where S_1 is a system term over a well-formed configuration $\Gamma c, \Pi \triangleright S_1$ in GR, then $\Gamma c, \Pi \triangleright S_1 \to {}^{k,p} \Gamma c, \Pi \triangleright S_2$ for some k and p such that $F(S_2) \equiv T_2'$ where $T_2' \equiv T_2$.

Therefore systems in both GR and PBL are proven to match up to structural equivalence under reduction semantics. Since PBL [17] is a specification for GR, therefore it has been shown that GR conforms to its specification. Further, the language PBL has well-established processes algebraic theory [17, 29]; therefore the choice of this model is justified to the extent of conformance with an established specification.

9 Conclusion

This model, GR, not only successfully demonstrates the quality of the services of the network based upon the number of hop crossings a delivery message has to take before getting it delivered to the waiting input process, but also this model can be used for performance analysis of the Industrial IoT System and the modern distributed network system based upon probability of the paths followed by message to deliver the value from process at source node to a process at the destination node. The entire setup of this model is also closer to real IoT's and distributed networks. Since PBL is a specification for GR, therefore we have shown that GR conforms to its specification. The language PBL has well-established process algebraic theory [17]; therefore our choice of this model is justified to the extent of reduction equivalence with its specification. Next, we intend to define a bisimulation-based preorder [6, 9] between configurations over a labelled transition system [6]. We wish to show that this preorder coincides with a relation defined as key observational properties over its reduction semantics.

References

1. J. Woodcock, P.G. Larsen, J. Bicarregui, J. Fitzgerald, Formal methods: Practice and experience. ACM Comput. Surv. **41**(4), 19:1–19:36 (2009)
2. R. Milner, *Communicating and mobile systems: The π-Calculus* (Cambridge University Press, 1999)
3. A. Souri, M. Norouzi, A state-of-the-art survey on formal verification of the internet of things applications. J. Serv. Sci. Res. **11**(1), 47–67 (2019)
4. M. Houimli, L. Kahloul, S. Benaoun, Formal specification, verification and evaluation of the MQTT protocol in the internet of things, in *2017 International Conference on Mathematics and Information Technology (ICMIT)*, (IEEE, 2017), pp. 214–221
5. S. Ouchani, Ensuring the functional correctness of IoT through formal modeling and verification, in *International Conference on Model and Data Engineering*, (Springer, 2018), pp. 401–417
6. M. Hennessy, *A distributed Pi-Calculus* (Cambridge University Press, 2007)
7. D. Sangiorgi, *An Introduction to Bisimulation and Coinduction* (Cambridge University Press, Cambridge/New York, 2012)
8. D. Sangiorgi, On the origins of bisimulation and coinduction. ACM Trans. Program. Lang. Syst. **31**(4), 15:1–15:41 (2009). https://doi.org/10.1145/1516507.1516510
9. M. Gaur: A Routing Calculus: Towards Formalising the Cost of Computation in a Distributed Computer Network. Phd, Informatics, University of Sussex, U.K. (December 2008)
10. Gaur, M.: A routing calculus for distributed computing. In Elena Troubitsyna, Proceedings of Doctoral Symposium held in conjunction with Formal Methods 2008,Turku Centre for Computer Science General Publication 48, 23–32. (2008)
11. M. Stolpe, The internet of things: Opportunities and challenges for distributed data analysis. ACM SIGKDD Explor. Newslett. **18**(1), 15–34 (2016)
12. P. Sethi, S.R. Sarangi, Internet of things: Architectures, protocols, and applications. Journal of Electrical and Computer Engineering **2017** (2017)
13. N. Sfeir, H. Sharifi: Internet of things solutions in smart cities

14. A.S. Tanenbaum, M. Van Steen, Distributed systems: Principles and paradigms. Prentice-Hall (2007)
15. K. Iwanicki, A distributed systems perspective on industrial IoT, in *2018 IEEE 38th International Conference on Distributed Computing Systems (ICDCS)*, (IEEE, 2018), pp. 1164–1170
16. M. Gaur, R. Kant, A survey on process algebraic stochastic modelling of large distributed systems for its performance analysis, in *3rd International Conference on Eco-friendly Computing and Communication Systems (ICECCS)*, (2014), pp. 206–211. https://doi.org/10.1109/Eco-friendly.2014.49
17. Cerone, A, Foundations of Ad Hoc Wireless Networks. Ph.D. thesis, School of Computer Science and Statistics, Trinity College Dublin (July 2012), https://www.doc.ic.ac.uk/acerone/works/thesis.pdf
18. G. Norman, D. Parker, *Quantitative verification: Formal guarantees for timeliness, reliability and performance*, Tech. rep. (The London Mathematical Society and the Smith Institute, 2014)
19. C. Huitema, *Routing in the Internet* (Prentice-Hall, Inc., Upper Saddle River, 1995)
20. J.F. Kurose, K.W. Ross, *Computer Networking: A Top-Down Approach Featuring the Internet* (Pearson Benjamin Cummings, 2004)
21. D.S.J. De Couto, D. Aguayo, B.A. Chambers, R. Morris, Performance of multihop wireless networks: Shortest path is not enough. SIGCOMM Comput. Commun. Rev. **33**(1), 83–88 (Jan 2003). https://doi.org/10.1145/774763.774776
22. Shukla, S.: An Adaptive Probabilistic Routing Algorithm. Ph.D. thesis, Department of Electrical Engineering, Indian Institute of Technology, Kanpur (May 2005), home.iitk.ac.in/˜ynsingh/mtech/swapnil.pdf
23. T.K. Huang, C.K. Lee, L.J. Chen, Prophet+: An adaptive prophet-based routing protocol for opportunistic network, in *2010 24th IEEE International Conference on Advanced Information Networking and Applications*, (2010), pp. 112–119. https://doi.org/10.1109/AINA.2010.162
24. J.S. Baras, H. Mehta, A Probabilistic Emergent Routing Algorithm for Mobile Ad Hoc Networks, in *WiOpt'03: Modeling and Optimization in Mobile, Ad Hoc and Wireless Networks*, (Sophia Antipolis, France, 2003) p. 10 pages, https://hal.inria.fr/inria-00466600
25. E. Curran, J. Dowling, Sample: Statistical network link modelling in an on-demand probabilistic routing protocol for ad hoc networks, in *Second Annual Conference on Wireless On-demand Network Systems and Services*, (2005), pp. 200–205. https://doi.org/10.1109/WONS.2005.30
26. K. Bhargavan, D. Obradovic, C.A. Gunter, Formal verification of standards for distance vector routing protocols. J. ACM **49**(4), 538–576 (2002). https://doi.org/10.1145/581771.581775
27. M. Tribastone, The pepa plug-in project, in *Proceedings of the Fourth International Conference on Quantitative Evaluation of Systems*, (QEST '07, IEEE Computer Society, Washington, DC, 2007), pp. 53–54. https://doi.org/10.1109/QEST.2007.49
28. M. Kwiatkowska, G. Norman, D. Parker, Prism: Probabilistic model checking for performance and reliability analysis. ACM SIGMETRICS Perform. Eval. Rev. **36**(4), 40–45 (2009)
29. M.H. Andrea Cerone, Modelling probabilistic wireless networks, in *Formal Techniques for Distributed Systems, Lecture Notes in Computer Science*, ed. by H. Giese, G. Rosu, vol. 7273, (Springer, Berlin/Heidelberg, 2012), pp. 135–151. https://doi.org/10.1007/978-3-642-30793-5_9
30. A. Cerone, *Matthew: A Simple Probabilistic Broadcast Language*, Tech. rep (Foundations and Methods Research Group, Trinity College Dublin, 2012)
31. A. Boukerche, S. Nikoletseas, Wireless communications systems and networks, in *Protocols for Data Propagation in Wireless Sensor Networks*, ed. by M. Guizani, (Plenum Press, New York, 2004), pp. 23–51. http://dl.acm.org/citation.cfm?id=1016648.1016651
32. K.S. Trivedi, *Probability and Statistics with Reliability, Queuing and Computer Science Applications*, 2nd edn. (Wiley, Chichester, 2002)

Formal Verification of Industrial IoT System: A Behavioural Equivalence

Rama Kant

1 Introduction

In recent years, formal verification [1], specifically in process algebraic framework [2], is widely used to model system behaviour and prove the correctness of IoT's [3–7] which are inherently distributed in nature and fit perfectly the classic definitions of distributed systems [8–12]. The stochastic extension of process algebras [13] and probabilistic extension of process algebras [14, 15] are developed to add quantification to process algebra models. We discuss behavioural equivalence of an stochastic extension of routing calculi [16, 17], *GR*, which is developed to model modern distributed systems in a better way by its implicit underlying formal distributed network architecture, also of industrial IoT system [8–11]. In process algebras [2, 18] framework, there are two semantic views of the underlying system. First semantic view is related to internal interactions, not seen by external entity, of the system namely called a reduction semantics [18]. The second semantic view is related to external interactions between the system and its surrounding environment namely called action semantics [18]. It interprets the language as a labeled transition system or lts [19, 20]. The labeled transition system has a very powerful underlying coinductive proof technique which is called bisimulation [19, 20]. It induces automatically a bisimulation equivalence between processes. We show that this pre-order coincides with a relation defined as key observational properties over its reduction semantics.

Funding: This work was supported by World Bank TEQIP-II project (2013–16) at Institute of Engineering and Technology, Lucknow, (Dr A P J Abdul Kalam Technical University, Lucknow, Uttar Pradesh, India).

R. Kant (✉)
Dr A P J Abdul Kalam Technical University, Lucknow, India

2   A Stochastic Routing Calculus

A typical system in, *GR*, will look like $\langle R^{\Pi} \rangle [\![n[P]]\!]$ with a process P at named node *n* which in turn is attached with a router *R*. The process P may communicate with another process resides at the same node or another node connected to some router. The communication between the processes takes place via routers. The routers determine the paths, along the router connectivity, through which the communicated values are propagated with some probability. Each router maintains a table, called a routing table represented as $\langle R^{\Pi} \rangle$, consisting of a mapping for each node name to all adjacent routers and probability of output process (i.e. message) along the communication link between routers which are on the paths of the destination node. There are three syntactic categories in this language. These are systems, nodes, and processes. The syntax of this calculus is given in Sect. 1 (Fig. 1).

Next, a formal relation between the systems, intuitively to represent the systems as same computational entities, is described, called structural equivalence denoted by $\equiv$ in *G*R and defined in a conventional way [2, 16, 18]. There are three different syntactic categories for system, nodes, and processes. The reduction semantics describe the allowable computations from individual processes and described with respect to the router connectivity and probability distribution over communication links between routers. It is defined as a binary relation between constructs called configurations. A configuration is defined in terms of network of routers and

Fig. 1 Syntax

$$S,\ T\ ::=$$

	Systems
$\langle R^{\Pi} \rangle [\![M]\!]$	Probabilistic Router
$S\mid T$	Concurrency
$[R]_{\boxtimes}^{k,P}\{n,m,v@c\}$	Messages
$(new\ d)\ S$	New Name
ε	Identity

$$M,\ N ::=$$

	Nodes
$n[P]$	Named Processes
$M\mid N$	Concurrency
$(new\ d)\ M$	New name
0	Identity

$$P,\ Q\ ::=$$

	Process Terms
$c?(x)\ P$	Input
$m!\langle v@c\rangle$	Output
$if\ v = u\ then\ P\ else\ Q$	Matching
$(new\ b)\ P$	Channel name creation
$newnode\ m\ with\ P\ in\ Q$	New Node Creation
$P\mid Q$	Concurrency
$m!\langle v@c\rangle\ _p \oplus m!\langle v@c\rangle$	Probabilistic Choice
$* P$	Repetition
$stop$	Identity
ω	Success

probability distribution. The network of routers Γc is basically a binary relation between the router names Figs. 2a, 2b and 3.

3 GR: A Behavioural Equivalence

Here, it is discussed that how we can say that one well-formed configuration, in *GR*, is more efficient than another one where both well-formed configurations exhibit the same behaviour. For this, a behavioural equivalence is defined for processes in the language, *GR*, on the basis of their different cost and probability of reduction. After that a probabilistic observational pre-order between well-formed configurations is defined, similar to [16, 18, 21], based upon three key probabilistic observational properties. This probabilistic observational pre-order is purely based on the reduction semantics of *GR* namely called probabilistic reduction barbed pre-congruence. Action semantics describes how processes in *GR* can interact with other processes by communicating along channels when placed in a large computational environment. This view of the language is represented by a labeled transition system or a lts [18–21]. Labeled transition system [19, 20] gives us a very powerful coinductive proof technique, which is called bisimulation equivalence, for establishing equivalences between processes [19, 20]. A bisimulation-based pre-order is defined which is called probabilistic efficiency pre-bisimulation based on a labeled transition system of *GR*. Further, observational labeled transition system for *GR* is given. This allows an observer to observe an action being performed by a well-formed configuration. This extension of labeled transition system is called

(R-OUT)
$$\Gamma_c, \Pi \vartriangleright \langle R^\Pi \rangle [\![n[m!\langle v@c \rangle \mid P] \mid N]\!] \longrightarrow \Gamma_c, \Pi \vartriangleright [R]_{\boxtimes}^{0,1} \{n, m, v@c\} \mid \langle R^\Pi \rangle [\![n[P] \mid N]\!]$$

(R-BCOM-I)
$$\frac{\begin{array}{l} \{(R_1, R_j)\} \in \Gamma_c \\ p_{1j} > 0, p_{1j} \in \Pi \\ \langle R_1^\Pi \rangle (m) = \{R_j\} \\ \langle R_j^\Pi \rangle (v) \uparrow, v \in \mathcal{NN} \end{array}}{\begin{array}{l} \Gamma_c, \Pi \vartriangleright [R_1]_{\boxtimes}^{k,P} \{n, m, v@c\} \mid \langle R_j^\Pi \rangle [\![N]\!] \mid S \longrightarrow \\ \Gamma_c, \Pi \vartriangleright [R_j]_{\boxtimes}^{k+1,P*p_{1j}} \{n, m, v@c\} \mid \langle R_j^\Pi \{v \to R_1\} \rangle [\![N]\!] \mid S \end{array}} \text{for all } j=2,3,..$$

(R-BCOM-II)
$$\frac{\begin{array}{l} \{(R_1, R_j)\} \in \Gamma_c \\ p_{1j} > 0, p_{1j} \in \Pi \\ \langle R_1^\Pi \rangle (m) = \{R_j\} \\ \langle R_j^\Pi \rangle (v) \downarrow, v \in \mathcal{NN} \end{array}}{\begin{array}{l} \Gamma_c, \Pi \vartriangleright [R_1]_{\boxtimes}^{k,P} \{n, m, v@c\} \mid \langle R_j^\Pi \rangle [\![N]\!] \mid S \longrightarrow \\ \Gamma_c, \Pi \vartriangleright [R_j]_{\boxtimes}^{k+1,P*p_{1j}} \{n, m, v@c\} \mid \langle R_j^\Pi \rangle [\![N]\!] \mid S \end{array}} \text{for all } j=2,3,..$$

Fig. 2a Reductions semantics (contd…)

(R-MSG–DEL)

$$\frac{\langle R^{\Pi}\rangle(m) = \{R\}}{\Gamma_c,\Pi \ \triangleright \ [R]_{\boxtimes}^{k,p}\{n,m,v@c\} \mid [R]_{\boxtimes}^{k',p'}\{n,m,v@c\} \mid S \longrightarrow}, if\ k < k'$$
$$\Gamma_c,\Pi \ \triangleright \ [R]_{\boxtimes}^{k,p}\{n,m,v@c\} \mid S$$

$$\frac{\langle R^{\Pi}\rangle(m) = \{R\}}{\Gamma_c,\Pi \ \triangleright \ [R]_{\boxtimes}^{k,p}\{n,m,v@c\} \mid [R]_{\boxtimes}^{k',p'}\{n,m,v@c\} \mid S \longrightarrow}, if\ k > k'$$
$$\Gamma_c,\Pi \ \triangleright \ [R]_{\boxtimes}^{k',p'}\{n,m,v@c\} \mid S$$

$$\frac{\langle R^{\Pi}\rangle(m) = \{R\}}{\Gamma_c,\Pi \ \triangleright \ [R]_{\boxtimes}^{k,p}\{n,m,v@c\} \mid [R]_{\boxtimes}^{k',p'}\{n,m,v@c\} \mid S \longrightarrow}, if\ k = k' and\ p > p'$$
$$\Gamma_c,\Pi \ \triangleright \ [R]_{\boxtimes}^{k,p}\{n,m,v@c\} \mid S$$

$$\frac{\langle R^{\Pi}\rangle(m) = \{R\}}{\Gamma_c,\Pi \ \triangleright \ [R]_{\boxtimes}^{k,p}\{n,m,v@c\} \mid [R]_{\boxtimes}^{k',p'}\{n,m,v@c\} \mid S \longrightarrow}, if\ k = k' and\ p < p'$$
$$\Gamma_c,\Pi \ \triangleright \ [R]_{\boxtimes}^{k',p'}\{n,m,v@c\} \mid S$$

$$\frac{\langle R^{\Pi}\rangle(m) = \{R\}}{\Gamma_c,\Pi \ \triangleright \ [R]_{\boxtimes}^{k,p}\{n,m,v@c\} \mid [R]_{\boxtimes}^{k',p'}\{n,m,v@c\} \mid S \longrightarrow}, if\ k = k' and\ p = p'$$
$$\Gamma_c,\Pi \ \triangleright \ [R]_{\boxtimes}^{k,p}\{n,m,v@c\} \mid S$$

(R-IN)

$$\frac{\langle R^{\Pi}\rangle(m) = \{R\},\ [R]_{\boxtimes}^{l,q}\{n,m,v@c\}\ \textbf{notin}\ S\ for\ any\ l\ and\ q}{\Gamma_c,\Pi \ \triangleright \ [R]_{\boxtimes}^{k,p}\{n,m,v@c\} \mid \langle R^{\Pi}\rangle[\![m[c?(x)\,P]\,]\!] \mid N]\!] \mid S \longrightarrow^{k,p}}$$
$$\Gamma_c,\Pi \ \triangleright \ \langle R^{\Pi}\rangle[\![m[P\{^v/x\}] \mid N]\!] \mid S$$

Fig. 2b Reductions semantics (contd…)

(R-NEWNODE–CREATION)
$$\Gamma_c,\Pi \ \triangleright \ \langle R^{\Pi}\rangle[\![n[newnode\ m\ with\ P\ in\ Q]]\!] \longrightarrow$$
$$\Gamma_c,\Pi \ \triangleright \ (new\ m)(\langle R^{\Pi}\{m \to R\}\rangle[\![m[P] \mid n[Q]]\!])$$

(R-MATCH)
$$\Gamma_c,\Pi \ \triangleright \ \langle R^{\Pi}\rangle[\![n[if\ v = u\ then\ P\ else\ Q]]\!] \longrightarrow \Gamma_c,\Pi \ \triangleright \ \langle R^{\Pi}\rangle[\![n[P]]\!]$$

(R-MISMATCH)
$$\Gamma_c,\Pi \ \triangleright \ \langle R^{\Pi}\rangle[\![n[if\ v_1 = v_2\ then\ P\ else\ Q]]\!] \longrightarrow \Gamma_c,\Pi \ \triangleright \ \langle R^{\Pi}\rangle[\![n[Q]]\!]\quad v_1 \neq v_2$$

(R-STRUCT)
$$\frac{S \equiv S',\Gamma_c,\Pi \ \triangleright \ S' \longrightarrow^{k,p} \Gamma_c,\Pi \ \triangleright \ R', R' \equiv R}{\Gamma_c,\Pi \ \triangleright \ S \longrightarrow^{k,p} \Gamma_c,\Pi \ \triangleright \ R}$$

(R-CNTX)
$$\frac{\Gamma_c,\Pi \ \triangleright \ S_1 \longrightarrow^{k,p} \Gamma_c,\Pi \ \triangleright \ S_1'}{\Gamma_c,\Pi \ \triangleright \ S_1 \mid S_2 \longrightarrow^{k,p} \Gamma_c,\Pi \ \triangleright \ S_1' \mid S_2}$$
$$\Gamma_c,\Pi \ \triangleright \ S_2 \mid S_1 \longrightarrow^{k,p} \Gamma_c,\Pi \ \triangleright \ S_2 \mid S_1'$$
$$\Gamma_c,\Pi \ \triangleright \ (new\ d)S_1 \longrightarrow^{k,p} \Gamma_c,\Pi \ \triangleright \ (new\ d)S_1'$$

Fig. 3 (Contd…) Reduction semantics

asynchronous labeled transition system. For this extended lts, asynchronous probabilistic efficiency pre-bisimulation is defined. After that, it is shown that both asynchronous probabilistic efficiency pre-bisimulation and probabilistic reduction barbed pre-congruence imply each other, thereby making the language fully abstract. Finally, it has shown that asynchronous probabilistic efficiency pre-bisimulation can be recovered from probabilistic reduction barbed pre-congruence and vice versa.

4 Probabilistic Reduction Barbed Pre-congruence

The motive is to develop a notion of probabilistic pre-order between the well-formed configuration $\Gamma c, \Pi_1 \triangleright S$ and $\Gamma c, \Pi_2 \triangleright T$. Now one shall define the properties which can be used to relate two well-formed configurations. Let us now define an observation in GR:

Definition 1. (Probabilistic observation) $\Gamma c, \Pi \triangleright S \downarrow^{k,p} [R]_{\boxtimes} \{n, m, v @ c\}$ if $S \equiv (new\ b)([R]_{\boxtimes}^{k,p} \{n,m,v@c\} \mid S')$ for some S' such that $\langle R^{\Pi} \rangle (m) = \{R\}$ where $c \neq b$.

$\Gamma c, \Pi \triangleright S \Downarrow^{k,p} [R]_{\boxtimes} \{n, m, v @ c\}$ can also be inductively defined as

Definition 2 $\Gamma c, \Pi \triangleright S \Downarrow^{k,p} [R]_{\boxtimes} \{n, m, v @ c\}$ if

1. $\Gamma c, \Pi \triangleright S \downarrow^{k,p} [R]_{\boxtimes} \{n, m, v @ c.\}$
2. if $\Gamma c, \Pi \triangleright S \xrightarrow{*}{}^{l,q} \Gamma c, \Pi \triangleright T$ then $\Gamma c, \Pi \triangleright T \Downarrow^{k-l, p/q} [R]_{\boxtimes} \{n, m, v @ c\}$

A relation between well-formed configurations called as probabilistic observation cost improving is defined as follows:

Definition 3. (Probabilistic observation cost improving) A binary relation R on well-formed configurations is probabilistic observation cost improving, if $(\Gamma c, \Pi_1 \triangleright S)R\ (\Gamma c, \Pi_2 \triangleright T)$ and $\Gamma c, \Pi_1 \triangleright S \Downarrow^{k,p} [R]_{\boxtimes} \{n, m, v @ c\}$ implies $\Gamma c, \Pi_2 \triangleright T \Downarrow^{l,q} [R]_{\boxtimes} \{n, m, v @ c\}$ for some l and q such that *either $l < k_l$ or $l = k$ and $q > p$*. The contextual property preserves the parallel composition of well-formed configurations. So, first define contextual relation in Definition 4.

Definition 4. (Contextual) A binary relation R on well-formed configurations is contextual.

1. If $S \equiv (new\ \tilde{k1})(\langle R^{\Pi 1} \rangle m[P_1] | N_1] | S_1)$, $T \equiv (new\ \tilde{k2})(\langle R^{\Pi 2} \rangle m[P_2] | N_2] | S_2)$ and for any Q such that $\Gamma c, \Pi 1 \triangleright (new\ \tilde{k1})(\langle R^{\Pi 1} \rangle m[P_1] | N_1] | S_1)$ and $\Gamma c, \Pi 2 \triangleright (new\ \tilde{k2})(\langle R^{\Pi 2} \rangle m[P_2] | N_2] | S_2)$ are well formed and $bn(Q) \notin fn(S, T)$ then $(\Gamma c, \Pi 1 \triangleright S)R(\Gamma c, \Pi 2 \triangleright T)$ implies $(\Gamma c, \Pi 1 \triangleright (new\ \tilde{k1})(\langle R^{\Pi 1} \rangle m[P_1] | N_1] | S_1))\ R\ (\Gamma c, \Pi 2 \triangleright (new\ \tilde{k2})(\langle R^{\Pi 2} \rangle m[P_2] | N_2] | S_2))$
2. If for messages, $[R]_{\boxtimes}^{k,p} \{n,m,v@c\}$ and $[R]_{\boxtimes}^{l,q} \{n,m,v@c\}$ such that $\langle R^{\Pi 1} \rangle (m) = R$, $\langle R^{\Pi 2} \rangle (m) = R$ where $\langle R^{\Pi 1} \rangle \in S$ and $\langle R^{\Pi 2} \rangle \in T$. Now, $\Gamma c, \Pi 1 \triangleright S | [R]_{\boxtimes}^{k,p} \{n,m,v@c\}$ with cost k and probability p, $\Gamma c, \Pi 2 \triangleright T | [R]_{\boxtimes}^{l,q} \{n,m,v@c\}$ with cost l and probability q are well formed. Then $(\Gamma c, \Pi 1 \triangleright S)\ R\ (\Gamma c, \Pi 2 \triangleright T)$ implies $(\Gamma c, \Pi 1 \triangleright S | [R]_{\boxtimes}^{k,p} \{n,m,v@c\})\ R\ (\Gamma c, \Pi 2 \triangleright T | [R]_{\boxtimes}^{l,q} \{n,m,v@c\})$.

Note that the definition of contextual relation is different than the conventional one [2] because the inputs are performed by processes whereas the output is done by messages which are created and propagated at the system level in three level syntactic construct in GN. We shall try to see if the second property of the contextual relation is satisfied in the structural equivalence in the following proposition:

Proposition 1. If $S \equiv T$ and $\Gamma c, \Pi \vartriangleright S$, $\Gamma c, \Pi \vartriangleright T$ are well formed, then for messages $\left[R \right]_{\boxtimes}^{k,p} \{n,m,v@c\}$ and $\left[R \right]_{\boxtimes}^{l,q} \{n,m,v@c\}$ such that $\langle\ R^{\Pi}\ \rangle(m) = R$ and $(\Gamma c, \Pi 1\ \vartriangleright S)$ $\left[R \right]_{\boxtimes}^{k,p} \{n,m,v@c\})$ and $(\Gamma c,\ \Pi 2 \vartriangleright T)\left[R \right]_{\boxtimes}^{l,q} \{n,m,v@c\})$ are well formed only if $k=l$ and $p=q$. Probabilistic reduction cost improving should take the internal branching structure of the system into account, defined as follows:

Definition 5. (Probabilistic reduction cost improving) A binary relation R on well-formed configuration is probabilistic reduction cost improving if $(\Gamma c, \Pi 1 \vartriangleright S)$ $R\ (\Gamma c, \Pi 2 \vartriangleright T)$ means

1. $\Gamma c, \Pi 1 \vartriangleright S \relbar\joinrel\rightarrow^{k,p} \Gamma c, \Pi 1 \vartriangleright S'$ implies $\Gamma c, \Pi 2 \vartriangleright T \relbar\joinrel\rightarrow^{l,q} \Gamma c, \Pi 2 \vartriangleright T'$ such that either $l < k$ or $l = k$ and $q > p$ and $(\Gamma c, \Pi 1 \vartriangleright S')R(\Gamma c, \Pi 2 \vartriangleright T')$.
2. $\Gamma c, \Pi 2 \vartriangleright T \relbar\joinrel\rightarrow^{l,q} \Gamma c, \Pi 2 \vartriangleright T'$ implies $\Gamma c, \Pi 1 \vartriangleright S \relbar\joinrel\rightarrow^{k,p} \Gamma c, \Pi 1 \vartriangleright S'$ such that either $l < k$ or $l = k$ and $q > p$ and $(\Gamma c, \Pi 1 \vartriangleright S')R(\Gamma c, \Pi 2 \vartriangleright T')$.

On the basis of the above three observational properties, now we are ready to define our touchstone equivalence by combining all these properties as follows:

Definition 6. (Probabilistic reduction barbed pre-congruence) A relation R is probabilistic reduction barbed pre-congruence if it is

- Probabilistic observation cost improving.
- Contextual.
- Probabilistic reduction cost improving.

$\preccurlyeq_{rbc}^{p}$ is the largest such relation.

We take $\preccurlyeq_{rbc}^{p}$ to be our touchstone equivalence, motivated by minimal set of desirable properties one might require of a semantic equivalence. It is also possible to use the definition of $\preccurlyeq_{rbc}^{p}$ in order to reason about it. This can be formally stated in the following lemma:

Lemma 1. For any system S and T, $S \equiv T$ implies $\Gamma c, \Pi 1 \vartriangleright S \preccurlyeq_{rbc}^{p} \Gamma c, \Pi 2 \vartriangleright T$ for well-formed configurations $\Gamma c, \Pi 1 \vartriangleright S$ and $\Gamma c, \Pi 2 \vartriangleright T$.

Now, a different view of configurations of GR is defined in Sect. 5, describing how they would behave when placed in a large computational environment consisting of other configurations.

5   A Labeled Transition System

How a configuration interacts with other configurations in a large computational environment? The possible interaction between a configuration and a large environment consists of creation, propagation, and delivery of messages, although the configuration may also decide to ignore the environment. Consequently this view of the configurations may be given in terms of a labeled transition systems (lts) [2, 18] where actions represent its ability to create, propagate, and deliver messages or compute internally. Since we define action judgements on well-formed configurations, the inference of those actions which are result of actions performed by two or more components of the well-formed configuration may not be straightforward. We use the notation N^ϕ to denote an abstract representation of a system N where N^ϕ contains all the routing tables of N with empty processes in its nodes.

For systems M and N, the notation $M^{M|N\phi}$ is used to represent $M^{M|N\phi} \equiv M|N^\phi$. Similarly, notation $N^{M\phi|N}$ is used to represent $N^{M\phi|N} \equiv M^\phi|N$. Decomposition and composition are formally defined in Definitions 7 and 8.

Definition 7. (Decomposition)
We define decomposition of a well-formed configuration $\Gamma c, \Pi \triangleright M| N$ into $\Gamma c, \Pi \triangleright M^{M|N\phi}$ and $\Gamma c, \Pi \triangleright M^{M\phi|N}$ where both $\Gamma c, \Pi \triangleright M^{M|N\phi}$ and $\Gamma c, \Pi \triangleright M^{M\phi|N}$ are well formed.

Definition 8. (Composition) We define a parallel composition of two well-formed configurations $\Gamma c, \Pi \triangleright M^{M|N\phi}$ and $\Gamma c, \Pi \triangleright M^{M\phi|N}$ where both $\Gamma c, \Pi \triangleright M^{M|N\phi}$ to a well-formed configuration $\Gamma c, \Pi \triangleright M| N$.

Before describing a lts for GR, we will define two functions δ and ρ which will be useful in defining this lts.

Definition 9. (δ) In a well-formed configuration $\Gamma c, \Pi \triangleright S$ where $R_1, R_2, R_3, ..., R_k$ in S and $\mathbf{path}(n,m) = R_1 \rightsquigarrow R_1 \rightsquigarrow ... \rightsquigarrow R_k$ such that $\langle R_1^\Pi \rangle(n) = \{R_1\}, \langle R_k^\Pi \rangle(m) = \{R_k\}$ and $\langle R_1^\Pi \rangle(m) = \{R_2\}, \langle R_2^\Pi \rangle(m) = \{R_3\}, ...\langle R_{k-1}^\Pi \rangle(m) = \{R_k\}$. We define a function δ: $NN \times NN- \rightarrow R$ where R is the set of real numbers as $\delta(n,m) = |\mathrm{path}(n,m)| - 1$.

So, function $\delta(n,m)$ gives the cost of the path which is the number of hops the message has travelled in a well-formed configuration $\Gamma c, \Pi \triangleright S$.

Definition 10. (ρ) In a well-formed configuration $\Gamma c, \Pi \triangleright S$ where $R_1, R_2, R_3, ..., R_k$ in S and $\mathbf{path}(n,m) = R_1 \rightsquigarrow R_1 \rightsquigarrow ... \rightsquigarrow R_k$ such that $\langle R_1^\Pi \rangle(n) = \{R_1\}, \langle R_k^\Pi \rangle(m) = \{R_k\}$ and $\langle R_1^\Pi \rangle(m) = \{R_2\}, \langle R_2^\Pi \rangle(m) = \{R_3\}, ...\langle R_{k-1}^\Pi \rangle(m) = \{R_k\}$.

We define a function ρ: $NN \times NN- \rightarrow [0,1]$ as $\rho(n,m) = p_{12}*p_{23}*...*p_{(k-1)k}$ where $(i,j,p_{ij}) \in \Pi$ for i = 1,2,3...and j = 1,2,3,...means the output process (i.e. message) hops with probability p_{ij} along the communication link between routers R_i and R_j. In above Definition, function $\rho(n,m)$ gives the probability of the path, the message has travelled in $\Gamma c, \Pi \triangleright S$, which is the production of probabilities of hops because the probabilities $p_{12}, p_{23}, ..., i_{(k-1)k}$ are independent [22].

Now we describe a lts for GR, as defined in Figs. 4 and 5. Now we discuss the lts rules one by one given in Sect. 4 and 5. The first rule (l-in) means that input action is only possible when the substitution of v for x in P is well defined. In case, if input value is a node name, then it should be in the domain of routing table which is given by third premise. If input value is a new node name, then this condition is ensured by the lts of GR by the use of backward learning update. In (l-in), the transition

(L-IN)

$\delta(n,m) = k$

$\rho(n,m) = p$

$\langle R^{\Pi} \rangle(v) \downarrow, v \in \mathcal{NN}$

$$\overline{\Gamma_c, \Pi \;\triangleright\; \langle R^{\Pi} \rangle \llbracket m[c?(x)\,P] \mid N \rrbracket \mid S \xrightarrow{R.n.m.c?(v)}{}^{k,p} \Gamma_c, \Pi \;\triangleright\; \langle R^{\Pi} \rangle \llbracket m[P\{v/x\}] \mid N \rrbracket \mid S}$$

(L-OUT)

$$\Gamma_c, \Pi \;\triangleright\; \langle R^{\Pi} \rangle \llbracket n[m!\langle v@c \rangle \mid P] \mid N \rrbracket \xrightarrow{\tau}{}^{0,1} \Gamma_c, \Pi \;\triangleright\; [R]_{\boxtimes}^{0,1}\{n,m,v@c\} \mid \langle R^{\Pi} \rangle \llbracket n[P] \mid N \rrbracket$$

(L-COM-I)

$\{(R_1, R_j)\} \in \Gamma_c$

$p_{1j} > 0, p_{1j} \in \Pi$

$\langle R_1^{\Pi} \rangle(m) = \{R_j\}$

$\langle R_j^{\Pi} \rangle(v) \uparrow, v \in \mathcal{NN}$

$$\overline{\Gamma_c, \Pi \;\triangleright\; [R_1]_{\boxtimes}^{k,p}\{n,m,v@c\} \mid \langle R_j^{\Pi} \rangle \llbracket N \rrbracket \mid S \xrightarrow{\tau}{}^{0,1} \Gamma_c, \Pi \;\triangleright\; [R_j]_{\boxtimes}^{k+1,p*p_{1j}}\{n,m,v@c\} \mid} \quad \text{for all } j=2,3,..$$

$\langle R_j^{\Pi}\{v \to R_1\} \rangle \llbracket N \rrbracket \mid S$

(L-COM-II)

$\{(R_1, R_j)\} \in \Gamma_c$

$p_{1j} > 0, p_{1j} \in \Pi$

$\langle R_1^{\Pi} \rangle(m) = \{R_j\}$

$\langle R_j^{\Pi} \rangle(v) \downarrow, v \in \mathcal{NN}$

$$\overline{\Gamma_c, \Pi \;\triangleright\; [R_1]_{\boxtimes}^{k,p}\{n,m,v@c\} \mid \langle R_j^{\Pi} \rangle \llbracket N \rrbracket \mid S \xrightarrow{\tau}{}^{0,1} \Gamma_c, \Pi \;\triangleright\; [R_j]_{\boxtimes}^{k+1,p*p_{1j}}\{n,m,v@c\} \mid} \quad \text{for all } j=2,3,..$$

$\langle R_j^{\Pi} \rangle \llbracket N \rrbracket \mid S$

(L-DELIVERY)

$\langle R^{\Pi} \rangle(m) = \{R\}, [R]_{\boxtimes}^{l,p}\{n,m,v@c\}$ **notin** S for any l

$$\overline{\Gamma_c, \Pi \;\triangleright\; [R]_{\boxtimes}^{k,p}\{n,m,v@c\} \mid S \xrightarrow{[R]_{\boxtimes}\{n,m,v@c\}}{}^{k,p} \Gamma_c, \Pi \;\triangleright\; \varepsilon \mid S}$$

(L-COMM)

$$\Gamma_c, \Pi \;\triangleright\; M^{M|N^{\phi}} \xrightarrow{R.n.m.c?(v)}{}^{k,p} \Gamma_c, \Pi \;\triangleright\; M_1^{M_1|N^{\phi}}, \Gamma_c, \Pi \;\triangleright\; N^{M^{\phi}|N} \xrightarrow{[R]_{\boxtimes}\{n,m,v@c\}}{}^{k,p} \Gamma_c, \Pi \;\triangleright\; N_1^{M^{\phi}|N_1}$$

$$\overline{\Gamma_c, \Pi \;\triangleright\; M \mid N \xrightarrow{\tau}{}^{k,p} \Gamma_c, \Pi \;\triangleright\; (new\ b)(M_1 \mid N_1)}$$

$$\Gamma_c, \Pi \;\triangleright\; N^{M^{\phi}|N} \xrightarrow{[R]_{\boxtimes}\{n,m,v@c\}}{}^{k,p} \Gamma_c, \Pi \;\triangleright\; N_1^{M^{\phi}|N_1}, \Gamma_c, \Pi \;\triangleright\; M^{M|N^{\phi}} \xrightarrow{R.n.m.c?(v)}{}^{k,p} \Gamma_c, \Pi \;\triangleright\; M_1^{M_1|N^{\phi}}$$

$$\overline{\Gamma_c, \Pi \;\triangleright\; M \mid N \xrightarrow{\tau}{}^{k,p} \Gamma_c, \Pi \;\triangleright\; (new\ b)(M_1 \mid N_1)}$$

(L-MATCH)

$$\Gamma_c, \Pi \;\triangleright\; \langle R^{\Pi} \rangle \llbracket n[if\ v = u\ then\ P\ else\ Q] \rrbracket \xrightarrow{\tau}{}^{0,1} \Gamma_c, \Pi \;\triangleright\; \langle R^{\Pi} \rangle \llbracket n[P] \rrbracket$$

(L-MISMATCH)

$$\Gamma_c, \Pi \;\triangleright\; \langle R^{\Pi} \rangle \llbracket n[if\ v_1 = v_2\ then\ P\ else\ Q] \rrbracket \xrightarrow{\tau}{}^{0,1} \Gamma_c, \Pi \;\triangleright\; \langle R^{\Pi} \rangle \llbracket n[Q] \rrbracket \quad v_1 \neq v_2$$

Fig. 4 Labeled transition system (LTS) (contd...)

takes place when a well-formed system has the capability to receive an input, along a channel which is located at a process at some node placed at router, from a process at some source node. The cost, k, and probability, p, of doing an input are easily computable by functions δ and ρ, respectively.

The lts rule (l-out) needs no explanation where an output process reside at node placed at router does an internal action and generates a message. The cost of this reduction is 0 and the probability is 1. The reduction rules, (l-com-i) and (l-com-ii), are used to propagate the message from one router to the other towards the destination node. Let us first consider the rule (l-com-i). As in this calculus the message hops to all adjacent routers with some probability and this information is given by routing table. Formally, this is given by first, second, and third premises. The

(L-MSG−DELETE)

$$\dfrac{\langle R^{\Pi}\rangle(m)=\{R\}}{\Gamma_c,\Pi \;\triangleright\; [R]_{\boxtimes}^{k,p}\{n,m,v@c\} \mid [R]_{\boxtimes}^{k',p'}\{n,m,v@c\} \mid S \xrightarrow{\tau\;0,1}}\; , if\; k < k'$$
$$\Gamma_c,\Pi \;\triangleright\; [R]_{\boxtimes}^{k,p}\{n,m,v@c\} \mid S$$

$$\dfrac{\langle R^{\Pi}\rangle(m)=\{R\}}{\Gamma_c,\Pi \;\triangleright\; [R]_{\boxtimes}^{k,p}\{n,m,v@c\} \mid [R]_{\boxtimes}^{k',p'}\{n,m,v@c\} \mid S \xrightarrow{\tau\;0,1}}\; , if\; k > k'$$
$$\Gamma_c,\Pi \;\triangleright\; [R]_{\boxtimes}^{k',p'}\{n,m,v@c\} \mid S$$

$$\dfrac{\langle R^{\Pi}\rangle(m)=\{R\}}{\Gamma_c,\Pi \;\triangleright\; [R]_{\boxtimes}^{k,p}\{n,m,v@c\} \mid [R]_{\boxtimes}^{k',p'}\{n,m,v@c\} \mid S \xrightarrow{\tau\;0,1}}\; , if\; k = k'\, and\; p > p'$$
$$\Gamma_c,\Pi \;\triangleright\; [R]_{\boxtimes}^{k,p}\{n,m,v@c\} \mid S$$

$$\dfrac{\langle R^{\Pi}\rangle(m)=\{R\}}{\Gamma_c,\Pi \;\triangleright\; [R]_{\boxtimes}^{k,p}\{n,m,v@c\} \mid [R]_{\boxtimes}^{k',p'}\{n,m,v@c\} \mid S \xrightarrow{\tau\;0,1}}\; , if\; k = k'\, and\; p < p'$$
$$\Gamma_c,\Pi \;\triangleright\; [R]_{\boxtimes}^{k',p'}\{n,m,v@c\} \mid S$$

$$\dfrac{\langle R^{\Pi}\rangle(m)=\{R\}}{\Gamma_c,\Pi \;\triangleright\; [R]_{\boxtimes}^{k,p}\{n,m,v@c\} \mid [R]_{\boxtimes}^{k',p'}\{n,m,v@c\} \mid S \xrightarrow{\tau\;0,1}}\; , if\; k = k'\, and\; p = p'$$
$$\Gamma_c,\Pi \;\triangleright\; [R]_{\boxtimes}^{k,p}\{n,m,v@c\} \mid S$$

(L-NEWNODE−CREATION)

$$\Gamma_c,\Pi \;\triangleright\; \langle R^{\Pi}\rangle[n[newnode\; m\; with\; P\; in\; Q]] \xrightarrow{\tau\;0,1}$$
$$\Gamma_c,\Pi \;\triangleright\; (new\; m)(\langle R^{\Pi}\{m \to R\}\rangle[m[P] \mid n[Q]])$$

(L-STRUCT)

$$\dfrac{S \equiv S',\Gamma_c,\Pi \;\triangleright\; S' \xrightarrow{\mu\;k,p} \Gamma_c,\Pi \;\triangleright\; R',R' \equiv R}{\Gamma_c,\Pi \;\triangleright\; S \xrightarrow{\mu\;k,p} \Gamma_c,\Pi \;\triangleright\; R}$$

(L-CNTX)

$$\dfrac{\Gamma_c,\Pi \;\triangleright\; S_1 \xrightarrow{\mu\;k,p} \Gamma_c,\Pi \;\triangleright\; S_1'}{\Gamma_c,\Pi \;\triangleright\; S_1 \mid S_2 \xrightarrow{\mu\;k,p} \Gamma_c,\Pi \;\triangleright\; S_1' \mid S_2}\; bn(\mu) \notin fn(N)$$
$$\Gamma_c,\Pi \;\triangleright\; S_2 \mid S_1 \xrightarrow{\mu\;k,p} \Gamma_c,\Pi \;\triangleright\; S_2 \mid S_1'$$

$$\dfrac{\Gamma_c,\Pi \;\triangleright\; S_1 \xrightarrow{\mu\;k,p} \Gamma_c,\Pi \;\triangleright\; S_1'}{\Gamma_c,\Pi \;\triangleright\; (new\; b)S_1 \xrightarrow{\mu\;k,p} \Gamma_c,\Pi \;\triangleright\; (new\; b)S_1'}\; (b) \notin n(\mu)$$

(L-OPEN)

$$\dfrac{\Gamma_c,\Pi \;\triangleright\; S \xrightarrow{\alpha_0\;k,p} \Gamma_c,\Pi \;\triangleright\; S'}{\Gamma_c,\Pi \;\triangleright\; (new\; b)S \xrightarrow{(b)\alpha_0\;k,p} \Gamma_c,\Pi \;\triangleright\; S'}\; b \notin bn(\alpha_0)$$

Fig. 5 (Contd…) Reduction semantics

information about adjacent routers is given by first premise. The second and third premises give next hop routers with respective probability. In this calculus if message is carrying simple value, then it simply forwarded towards destination. But the message is carrying node name and unknown to intermediate nodes; then routing tables of intermediate routers must be updated about this node information which is called backward learning updates. This is given by fourth premise. The probability distribution is not affected here as probability is not assigned to output process along the communication link between node and router. The probability is assigned to output process along the communication link between routers. If some router is on the way of propagation of new node name, therefore its routing table is updated. This is one way of updating the entries about the newly created nodes. This method of routing table update is known as backward learning. The cost and the probability of this reduction are 0 and 1, respectively.

The reduction rule (l-com-ii) is similar to (l-com-i) except that there is no routing table update because message is carrying node name but is known to intermediate routers. The reduction rule (l-msg-delete) is used to delete the duplicate messages upon receiving at router of the destination node. The message is delivered by single message only as all duplicate messages are deleted by (l-msg-delete). This is ensured by premises. In rule (l-delivery), the resultant configuration does an output action at the home router of destination node. The cost of this reduction is k with probability p. Also note that since the labeled transition system is defined on well-formed configurations, therefore routing table must appear in system. The rule (*l-comm*) consists of two symmetric rules and defines the communication rule as simultaneous action of input and output actions. There are number of properties on these reductions in the following proposition:

Proposition 2. (Sanity Checks)
- $M \equiv N$ implies $M^\phi \equiv N^\phi$
- $M^{M|N\phi} \equiv M^{M|N}{}_1{}^\phi$, if $N^\phi \equiv N_1{}^\phi$
- if $\Gamma c, \Pi \triangleright M$ $\Gamma c, \Pi \triangleright M_1$ then $M^\phi \equiv M_1{}^\phi$.

Now we will see that well-formedness of configurations is also preserved under labeled transition system. It is stated in the following lemma:

Lemma 2. (Well-formed lts reduction) If $\Gamma c, \Pi \triangleright S$ is a well-formed configuration and $\Gamma c, \Pi \triangleright S \xrightarrow{\mu\ k,p} \Gamma c, \Pi \triangleright S'$, then $\Gamma c, \Pi \triangleright S'$ is also well-formed.

Now we are ready to find relation between the internal action relation and reduction relation, and it is stated in the following lemma:

Lemma 3. (Lts reduction up to structural equivalence) If $S \equiv T$ and $\Gamma c, \Pi \triangleright S \xrightarrow{\mu\ k,p} \Gamma c, \Pi \triangleright S'$, then there exists T' such that $\Gamma c, \Pi \triangleright T \xrightarrow{\mu\ k,p} \Gamma c, \Pi \triangleright T'$ such that $S' \equiv T'$.

Proposition 3. If $\Gamma c, \Pi \triangleright M \xrightarrow{(b)[R]_{\boxtimes} \{n,m,v@c\}^{k,p}} \Gamma c, \Pi \triangleright M'$ and $\langle R^\Pi \rangle (m) = \{R\}$, then $M \equiv (new\ b)\ ([R]_{\boxtimes}^{k,p} \{n,m,v@c\} | M')$.

Proposition 4. If $\Gamma c, \Pi \triangleright M \xrightarrow{R.n.m.c?(v)^{k,p}} \Gamma c, \Pi \triangleright M'$, then $M \equiv (new\ b)\ (\langle R^\Pi \rangle\ [\![m[c?(x)P]\ |N]\!]) | S)$ and $M' \equiv (new\ b)\ (\langle R^\Pi \rangle\ [\![m[[P\{v/x\}]\ |N\ ]\!] | S)$, where $(b) \bigcap n(v) = \phi$, $\delta(n,\ m) = k$ and $\rho(n,\ m) = p$.

Now we also have following consequence of Proposition 3 and 4:

Corollary 1. $\Gamma c, \Pi \triangleright M \xrightarrow{(b)[R]_{\boxtimes} \{n,m,v@c\}^{k,p}} \Gamma c, \Pi \triangleright M'$ implies $\delta(n,m) = k$ and $\rho(n,m) = p$.

Now we are ready to state the following two lemmas to establish relationship between $\rightarrow$ and $\twoheadrightarrow$.

Lemma 4. $\Gamma c,\Pi \rhd M \xrightarrow{\tau \ k,p} \Gamma c,\Pi \rhd M'$ implies $\Gamma c,\Pi \rhd M \twoheadrightarrow^{k,p} \Gamma c,\Pi \rhd M'$.

Now in the following lemma, we prove the converse of Lemma 4.

Lemma 5. $\Gamma c,\Pi \rhd M \twoheadrightarrow^{k,p} \Gamma c,\Pi \rhd M'$ where $\Gamma c,\Pi \rhd M$ is a well-formed configura-tion implies $\Gamma c,\Pi \rhd M \xrightarrow{\tau \ k,p} \Gamma c,\Pi \rhd M''$ such that $M' \equiv M''$. Finally, it can be summed up that the relationship between and reduction relation, using Lemma 4 and 5:

Theorem 1. $\Gamma c,\Pi \rhd M \twoheadrightarrow^{k,p} \Gamma c,\Pi \rhd M'$ iff $\Gamma c,\Pi \rhd M \xrightarrow{\tau \ k,p} \Gamma c,\Pi \rhd M''$ such that $M' \equiv M''$.

The relations in lts are closed in well-formed configurations defined for GR. This in turn means that we can have a bisimulation relation between well-formed con-figurations which is defined in Sect. 6.

6 Probabilistic Efficiency Pre-Bisimulation

We have discussed the labeled transition systems(lts) given in Figs. 4 and 5, the rela-tions over well-formed configurations are given by the actions $\xrightarrow{\tau \ k,p}$. This in turn means that we can have a bi-simulation relation between well-formed configura-tions. We refer both input and output action by the symbol α. We denote μ to repre-sent both external action α and internal action τ. In GR, to relate two well-formed configurations $\Gamma c,\Pi_1 \rhd S$ and $\Gamma c,\Pi_2 \rhd T$ such that later is "at least as good as " the former, it is sufficient to show that if $\Gamma c,\Pi_1 \rhd S$ can perform a action in cost k with probability p then $\Gamma c,\Pi_2 \rhd T$ should be able to do the same weak action in cost l with probability q such that **either l $<$k or l = k and q $>$p** .

This bi-simulation is not symmetric, and we call it probabilistic efficiency pre-bisimulation. Formally it is defined as follows:

Definition 11. (Probabilistic efficiency pre-bisimulation) A binary relation R on well-formed configurations is probabilistic efficiency pre-bisimulation if $\Gamma c,\Pi_1 \rhd S$ R $\Gamma c,\Pi_2 \rhd T$, where $\Gamma c,\Pi_1 \rhd S$ and $\Gamma c,\Pi_2 \rhd T$ are well-formed configurations, implies that

1. $\Gamma c,\Pi_1 \rhd S \xrightarrow{\mu \ k,p} \Gamma c,\Pi_1 \rhd S'$ for some k and p, then there exist a T' such that $\Gamma c,\Pi_2 \rhd T \xrightarrow{\hat{\mu} \ l,q} \Gamma c,\Pi_2 \rhd T'$ for some l and q such that **either l $<$ k or l = k and q $>$ p** . Also, $(\Gamma c,\Pi_1 \rhd S')$ R $(\Gamma c,\Pi_2 \rhd T')$.

2. $\Gamma c,\Pi_2 \triangleright T \xrightarrow{\mu \; l,q} \Gamma c,\Pi_2 \triangleright T'$ for some l and q, then there exist a S′ such that $\Gamma c,\Pi_1$
$\triangleright S \xrightarrow{\hat{\mu} \; l,q} \Gamma c,\Pi_2 \triangleright S'$ for some k and p such that **either l < k or l = k and q > p** .
Also, $(\Gamma c,\Pi_1 \triangleright S')\; R\; (\Gamma c,\Pi_2 \triangleright T')$.

$\preceq^{pe}$ is such largest relation.

Reader can take any example from Fig. 6 and 7 (exercise for readers) as per previous chapter and can see that if there are multiple paths between pair of nodes and different routing table entries at each routing table. If a configuration which is more efficient may be less efficient in the reverse direction of communication path. This is the limitation of our definition of probabilistic efficiency pre-bisimulation. In fact, Lemma 3 provides a method of working with efficiency pre-bisimulations up to structural equivalence. Let us first see a simple consequence of this lemma

Corollary 2. In GR, $S \equiv T$ implies $\Gamma c,\Pi \triangleright S \preceq^{pe} \Gamma c,\Pi \triangleright T$ where $\Gamma c,\Pi \triangleright S$ and $\Gamma c,\Pi \triangleright T$ are well-formed configurations.

We shall now see that in the definition of probabilistic efficiency pre-bisimulation when matching the residuals of actions, it is sufficient to work up to structural equivalence in the following proposition:

Proposition 5. If $(\Gamma c,\Pi \triangleright S)\; R\; (\Gamma c,\Pi \triangleright T)$, where R is a probabilistic efficiency pre-bisimulation up to structural equivalence, then $(\Gamma c,\Pi \triangleright S) \preceq^{pe} (\Gamma c,\Pi \triangleright T)$. Here are some interesting lemmas on the properties of lts given in Figs. 4 and 5.

Lemma 6. In a well-formed configuration $\Gamma c,\Pi \triangleright (new\; \widetilde{k1}))(\langle\; R^\Pi\; \rangle\; [\![m[P]\; |N]\!])|S$ if $\Gamma c,\Pi \triangleright (new\; \widetilde{k1})\; (\langle\; R^\Pi\; \rangle\; [\![m[P]\; |N]\!])|S \xrightarrow{\mu \; k,p} \Gamma c,\Pi \triangleright (new\; \widetilde{k2})\; (\langle\; R^\Pi\; \rangle\; \rangle [\![n[P']\; |N']$ $]\!]|S')$, then for any Q such that $\Gamma c,\Pi \triangleright (new\; \widetilde{k1}))(\langle\; R^\Pi\; \rangle\; [\![m[P|Q]\; |N]\!])|S$ is well-formed, $\Gamma c,\Pi \triangleright (new\; \widetilde{k1}))(\langle\; R^\Pi\; \rangle [\![\; n[P|Q]|N]\!])|S \xrightarrow{\mu \; k,p} \Gamma c,\Pi \triangleright (new\; \widetilde{k2})\; (\langle\; R^\Pi\; \rangle\; \rangle$ $[\![n[P'|Q]\; |N'\;]\!]\;]|S')$.

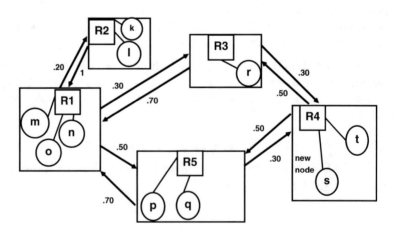

Fig. 6 A simple probabilistic distributed network with routers

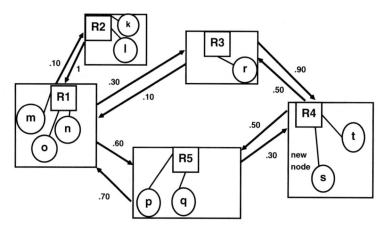

Fig. 7 A simple probabilistic distributed network with routers

Lemma 7. If a well-formed system $\Gamma c,\Pi \rhd (new\ \tilde{k})\ (\langle\ R^\Pi\ \rangle\ [\![m[P]\ |N]\!])|S$ does $\Gamma c,\Pi \rhd (new\ \tilde{k})\ (\langle\ R^\Pi\ \rangle\ [\![m[P]\ |N]\!])|S\ \overset{(b)[R]_\boxtimes\{n,m,b@c\}^{k,p}}{\rightarrow}\ \Gamma c,\Pi \rhd (new\ \tilde{k})\ (\langle\ R^\Pi\ \rangle\ [\![m[P]\ |N]\!])|S_1$, then for any Q such that $\Gamma c,\Pi \rhd (new\ \tilde{k})\ (\langle\ R^\Pi\ \rangle\ [\![m[P c?(x)Q]\ |N]\!])|S$ is well-formed $\Gamma c,\Pi \rhd (new\ \tilde{k})\ (\langle\ R^\Pi\ \rangle\ [\![m[P c?(x)Q]\ |N]\!])|S)\ \overset{\tau\ k,p}{\rightarrow}\ \Gamma c,\Pi \rhd (new\ \tilde{k},b)\ (\langle\ R^\Pi\ \rangle\ [\![m[P Q\{b/x\}]\ |N]\!])|\ S_1)$.

Corollary 3. For a well-formed configuration $\Gamma c,\Pi \rhd (new\ \widetilde{k1}))(\langle\ R^\Pi\ \rangle\ [\![m[P]\ |N]\!])|S$ and $\Gamma c,\Pi \rhd (new\ \widetilde{k1})\ (\langle\ R^\Pi\ \rangle\ [\![m[P]\ |N]\!])|S\ \overset{(b)[R]_\boxtimes\{n,m,b@c\}^{k,p}}{\rightarrow}\ \Gamma c,\Pi \rhd (new\ \widetilde{k2})$ $(\langle\ R^\Pi\ \rangle\ [\![m[P']\ |N']\!])|S'$ implies $\Gamma c,\Pi \rhd (new\ \tilde{k1})\ (\langle\ R^\Pi\ \rangle\ [\![m[P c?(x)Q]\ |N]\!])|S)\ \overset{\tau\ k,p}{\rightarrow}$ $\Gamma c,\Pi \rhd (new\ \widetilde{k2},b)\ (\langle\ R^\Pi\ \rangle\ [\![m[P'|Q\{b/x\}]\ |N']\!])|\ S')$ for any Q such that $\Gamma c,\Pi \rhd (new\ \tilde{k})\ (\langle\ R^\Pi\ \rangle\ [\![m[P c?(x)Q]\ |N]\!])|S$ is well-formed.

Proposition 6. In GR, The relation $\preceq^{pe}$ is contextual.

Proposition 7. In GR, the relation $\preceq^{pe}$ is probabilistic observation cost preserving.

Proposition 8. In GR, The relation $\preceq^{pe}$ is reduction cost improving.

Now from Proposition 6, 7, and 8, we conclude that $\preceq^{pe}$ preserves all the defining properties of $\precsim^{p}_{rbc}$. This is given is the following lemma:

Lemma 8. If $P \preceq^{pe} Q$ then $P \precsim^{p}_{rbc} Q$.

To prove the converse of Lemma 8, we first define an observational labeled transition system in Sect. 7.

7 An Observational Lts

Here we take the reduction semantics given in Figs. 2a, 2b and 3 in Sect. 2 and discuss the extent to which observers can see actions being performed by processes. We make an assumption that the input test process records the source node name, the cost, and probability of delivery for the delivered value v. We have to weaken the input rules used in Fig. 8.

Now we provide certain properties of new asynchronous input action in Lemma 9.

Lemma 9.

- $\Gamma c,\Pi \rhd M \xrightarrow{R.n.m.c?(v)^{k,p}} \Gamma c,\Pi \rhd M'$ implies $\Gamma c,\Pi \rhd M \xrightarrow{R.n.m.c?(v)^{k,p}}_a \Gamma c,\Pi \rhd M'$.

- $\Gamma c,\Pi \rhd M^{M|N\phi} \xrightarrow{R.n.m.c?(v)^{k,p}}_a \Gamma c,\Pi \rhd M_1{}^{M_1{}^{|N\phi}}, \Gamma c, \Pi \rhd M^{M\phi|N} \xrightarrow{R.n.m.c?(v)^{k,p}} \Gamma c,\Pi \rhd N_1$
  $^{M\phi|N_1}$ implies $\Gamma c,\Pi \rhd M|N \xrightarrow{\tau}{}^{k,p} \Gamma c,\Pi \rhd (new\ b)(M_1\ |N_1)$.

or

- $\Gamma c,\Pi \rhd M\ |N \equiv (new\ b)\ (M_1\ |N_1)$ where $\Gamma c,\Pi \rhd M\ |N$ and $(new\ b)\ (M_1\ |N_1)$ are well-formed configurations.

With these inference rules, we now have a new interpretation of GR as an lts:

- The configurations are closed terms in GR.

$$
\begin{array}{l}
\text{(L-A–IN)}\\
\delta(n,m) = k\\
\rho(n,m) = p\\
\langle R^\Pi \rangle(v) \downarrow, v \in \mathcal{NN}\\
\hline
\Gamma_c, \Pi \rhd \langle R^\Pi \rangle \llbracket m[c?(x)\ P]\ |\ N \rrbracket\ |\ S \xrightarrow{R.n.m.c?(v)}{}^{k,p} \Gamma_c, \Pi \rhd \langle R^\Pi \rangle \llbracket m[P\{v/x\}]\ |\ N \rrbracket\ |\ S
\end{array}
$$

$$
\begin{array}{l}
\text{(L-A–DELIVERY)}\\
\forall \langle R^\Pi \rangle(v) \downarrow, v \in \mathcal{NN}\\
\langle R^\Pi \rangle(m) = \{R\}\\
\delta(n,m) = k\\
\rho(n,m) = p\\
\hline
\Gamma_c, \Pi \rhd S \xrightarrow{R.n.m.c?(v)}{}^{0,1}_a \Gamma_c, \Pi \rhd [R]^{k,p}_{\boxtimes} \{n,m, v@c\}\ |\ S
\end{array}
$$

$$
\begin{array}{l}
\text{(L-A–CNTX)}\\
\dfrac{\Gamma_c, \Pi \rhd S_1 \xrightarrow{\mu}{}^{k,p} \Gamma_c, \Pi \rhd S_1'}{\Gamma_c, \Pi \rhd S_1\ |\ S_2 \xrightarrow{\mu}{}^{k,p} \Gamma_c, \Pi \rhd S_1'\ |\ S_2}\ \text{bn}(\mu) \notin \text{fn}(N)\\[2ex]
\Gamma_c, \Pi \rhd S_2\ |\ S_1 \xrightarrow{\mu}{}^{k,p} \Gamma_c, \Pi \rhd S_2\ |\ S_1'\\[2ex]
\dfrac{\Gamma_c, \Pi \rhd S_1 \xrightarrow{\mu}{}^{k,p} \Gamma_c, \Pi \rhd S_1'}{\Gamma_c, \Pi \rhd (new\ b)S_1 \xrightarrow{\mu}{}^{k,p} \Gamma_c, \Pi \rhd (new\ b)S_1'}\ (b) \notin \text{n}(\mu)
\end{array}
$$

Fig. 8 An Asynchronous Labeled Transition System

- The relations over well-formed configurations are given by actions.

  - $\Gamma c,\Pi \rhd S \xrightarrow[a]{R.n.m.c\,?(v)^{k,p}} \Gamma c,\Pi' \rhd S_1.$

  - $\Gamma c,\Pi \rhd S \xrightarrow[a]{(b)[R]_{\boxtimes}\{n,m,v@c\}^{k,p}} \Gamma c,\Pi' \rhd S_1.$

  - $\Gamma c,\Pi \rhd S \xrightarrow{\tau\ \ k,p} \Gamma c,\Pi' \rhd S_1.$

We refer to this as asynchronous lts for GR. For notational convenience, an arbitrary action in this lts will be denoted by $\Gamma c,\Pi \rhd S \xrightarrow[a]{\mu\ \ k,p} \Gamma c,\Pi' \rhd S_1$ although the output and internal actions will coincide with $\Gamma c,\Pi \rhd S \xrightarrow{\mu\ \ k,p} \Gamma c,\Pi' \rhd S_1$. It is worth pointing out that internal action $\Gamma c,\Pi \rhd S \xrightarrow{\tau}_a \Gamma c,\Pi' \rhd S1$ is still defined in terms of synchronous input and output actions.

We define a pre-bisimulation on asynchronous lts similar to that defined in Sect. 6.

Definition 12. (Asynchronous probabilistic efficiency pre-bisimulation) A binary relation R on well-formed configurations is asynchronous efficiency pre-bisimulation if $\Gamma c,\Pi \rhd S\ R\ \Gamma c,\Pi \rhd T$, where $\Gamma c,\Pi \rhd S$ and $\Gamma c,\Pi \rhd T$ are well-formed configurations, implies that

1. $\Gamma c,\Pi \rhd S \xrightarrow[a]{\mu\ \ k,p} \Gamma c,\Pi' \rhd S'$, then there exist a T' such that $\Gamma c,\Pi \rhd T \xRightarrow[a]{\mu\ \ k,p} \Gamma c,\Pi \rhd T'$ for some k, p, l and q such that **either l < k or l = k and q > p** and $(\Gamma c,\Pi \rhd S')\ R\ (\Gamma c,\Pi \rhd T')$.

2. $\Gamma c,\Pi \rhd T \xrightarrow[a]{\mu\ \ k,p} \Gamma c,\Pi' \rhd T'$, then there exist a S' such that $\Gamma c,\Pi \rhd S \xRightarrow[a]{\mu\ \ k,p} \Gamma c,\Pi \rhd S'$ for some k, p, l and q such that **either l < k or l = k and q > p** and $(\Gamma c,\Pi \rhd S')\ R\ (\Gamma c,\Pi \rhd T')$.

$\preceq_a^{pe}$ is the largest such relation. Now we describe certain properties of GR on asynchronous lts in line with properties described in Sect. 6.

Lemma 10. If $S \equiv T$ and $\Gamma c,\Pi \rhd S \xrightarrow[a]{\mu\ \ k,p} \Gamma c,\Pi \rhd S'$, then there exists some T' s.t. $\Gamma c,\Pi \rhd T \xrightarrow[a]{\mu\ \ k,p} \Gamma c,\Pi \rhd T'$ and $S' \equiv T'$.

Lemma 11. For any process Q if $\Gamma c,\Pi \rhd (new\ \widetilde{k}1))(\langle R^{\Pi} \rangle\ [\![n[P]\!|N]\!])|S$ and $\Gamma c,\Pi \rhd (new\ \widetilde{k}1))(\langle R^{\Pi} \rangle\ [\![n[P\!|Q]\!|N]\!])|S$ are well-formed and $\Gamma c,\Pi \rhd (new\ \widetilde{k}1))(\langle R^{\Pi} \rangle\ [\![n[P]\!|N]\!])|S \xrightarrow[a]{\mu\ \ k,p} \Gamma c,\Pi \rhd (new\ \widetilde{k}2))(\langle R^{\Pi} \rangle\ )[\![n[P]\ |N'\ ]\!]|S')$, then $\Gamma c,\Pi \rhd (new\ \widetilde{k}1))(\langle R^{\Pi} \rangle [\![\ n[P\!|Q]\ |N]\!])|S \xrightarrow[a]{\mu\ \ k,p} \Gamma c,\Pi \rhd (new\ \widetilde{k}2))(\langle R^{\Pi} \rangle [\![n[P'\!|Q]\ |N'\ ]\!]|S')$.

Corollary 4. In GR, $S \equiv T$ implies $\Gamma c,\Pi \rhd S \preceq_a^{pe} \Gamma c,\Pi \rhd T$.

Proposition 9. If $\Gamma c,\Pi \rhd SR\Gamma c,\Pi \rhd T$, where R is an asynchronous probabilistic efficiency pre-bisimulation up to structural equivalence, then $\Gamma c,\Pi \rhd S \preceq pe_a \Gamma c,\Pi \rhd T$.

8 Justifying Asynchronous Probabilistic Efficiency Pre-bisimulation Contextually

First we prove that $\preceq_a^{pe}$ preserves all the three defining properties, i.e. contextual, probabilistic observation cost improving, and probabilistic reduction cost improving in the following propositions.

Proposition 10 In GR, the relation $\preceq_a^{pe}$ is contextual.

Proposition 11 In GR, the relation $\preceq_a^{pe}$ is probabilistic observation cost preserving.

Proposition 12 In GR, the relation $\preceq_a^{pe}$ is reduction cost preserving.

$\preceq_a^{pe}$ implies $\preccurlyeq_{rbc}^{p}$ is given in the following lemma.

Lemma 12 If $\Gamma c,\Pi \, \triangleright S \preceq_a^{pe} \Gamma c,\Pi \, \triangleright T$, then $\Gamma c,\Pi \, \triangleright S \preccurlyeq_{rbc}^{p} \Gamma c,\Pi \, \triangleright T$.

Proof From Proposition 10, 11, and 12.

To prove the converse, we need to show that for every asynchronous action, there is a context which can observe it. We define a partial function *puts* which basically composes a test to a well-formed configuration according to the definition of contextual relation defined in Definition 13. This will be useful in proving that for every input and output action there is a context which can observe it.

Definition 13. (Contextual operator contextual operatorputs)
We define a partial function puts $C \times S \rightharpoonup C$ where C is a set of well-formed configuration and S is set of systems, as

1. If $\Gamma c,\Pi \, \triangleright S \equiv \Gamma c,\Pi \, \triangleright (newb)(\langle R^{\Pi}\rangle \, n[P]\|N\|)|S'$ is well-formed and for any $\langle R^{\Pi}\rangle$ $[\![n[Q]]\!]$ such that $\Gamma c,\Pi \, \triangleright (newb)(\, \langle R^{\Pi}\rangle n[P Q]\|N \, \|)|S'$ is well-formed, then $puts(\Gamma c,\Pi \, \triangleright S, \, \langle R^{\Pi}\rangle \, [\![n[Q]]\!]) = \Gamma c,\Pi \, \triangleright (newb)(\, \langle R^{\Pi}\rangle \, n[P Q]\|N \, \|)|S'$.
2. For any $\left[R\right]_{\boxtimes}^{k,p}\{n,m,v@c\}$ and a well-formed $\Gamma c,\Pi \, \triangleright S$ such that $\langle R^{\Pi}\rangle(R)=\{R\}$, $\delta(n,\ m)= k$, $\rho(n,m) = p$ and $\Gamma c,\Pi \, \triangleright S \, [\![R]\!]_{\boxtimes}^{k,p}\{n,m,v@c\}$ is well-formed, then $puts(\Gamma c,\Pi \, \triangleright S \,,\left[R\right]_{\boxtimes}^{k,p}\{n,m,v@c\}) = \Gamma c,\Pi \, \triangleright S \,|\left[R\right]_{\boxtimes}^{k,p}\{n,m,v@c\}$.
3. For a well-formed configuration $\Gamma c,\Pi \, \triangleright S$ with $T_1|T_2$ where T_i is a message, then $puts(\Gamma c,\Pi_1 \, \triangleright S_1, \, T_1|T_2) = puts(\, (puts(\Gamma c,\Pi \, \triangleright S, \, T_1), \, T_2)$.

Lemma 13.
1. If $\Gamma c,\Pi_1 \, \triangleright S_1 \preccurlyeq_{rbc}^{p} \Gamma c,\Pi_2 \, \triangleright S_2$ and $S_1 \equiv (newb)(\, \langle R^{\Pi 1}\rangle \, [\![n[P_1]\|N_1 \,]\!])|S_1'$, $S_2 \equiv (newb)(\, \langle R^{\Pi 2}\rangle [\![\, n[P_2]\|N_2 \,]\!])|S_2'$ such that for any $\langle R^{\Pi}\rangle [\![n[P]]\!]$, $puts(\Gamma c,\Pi_1 \, \triangleright S_1, \langle R^{\Pi}\rangle \, [\![n[P]]\!])$ and $puts(\Gamma c,\Pi_2 \, \triangleright S_2, \, \langle R^{\Pi}\rangle \, [\![n[P]]\!])$ are well defined, then $puts(\Gamma c,\Pi_1 \, \triangleright S_1, \, \langle R^{\Pi}\rangle \, [\![n[P]]\!]) \preccurlyeq_{rbc}^{p} puts(\Gamma c,\Pi_2 \, \triangleright S_2, \, \langle R^{\Pi}\rangle \, [\![n[P]]\!])$.
2. $\Gamma c,\Pi_1 \, \triangleright S_1 \preccurlyeq_{rbc}^{p} \Gamma c,\Pi_2 \, \triangleright S_2$ and for messages of the form $\left[R\right]_{\boxtimes}^{k,p}\{n,m,v@c\}, \left[R\right]_{\boxtimes}^{l,q}$ $\{n,m,v@c\}$ where $puts(S_1 \,, \left[R\right]_{\boxtimes}^{k,p}\{n,m,v@c\})$ and $puts(S_2, \left[R\right]_{\boxtimes}^{l,q}\{n,m,v@c\})$ are well defined and $\delta(n,\ m)= k$, $\rho(n,m) = p \in \Gamma c,\Pi_1 \, \triangleright S_1$, $\delta(n,m) = l$, $\rho(n,m)=q \in$

$\Gamma c, \Pi_2 \rhd S_2$, then $puts(\Gamma c, \Pi_1 \rhd S_1, [R]_{\boxtimes}^{k,p}\{n,m,v@c\}) \precsim_{rbc}^{p} puts(\Gamma c, \Pi_2 \rhd S_2, [R]_{\boxtimes}^{l,q}\{n,m,v@c\})$.

3. If $\Gamma c, \Pi_1 \rhd S_1 \precsim_{rbc}^{p} \Gamma c, \Pi_2 \rhd S_2$, then for any $T_1|T_2$ and $T_3|T_4$, where T_i is a message, such that $puts(\Gamma c, \Pi_1 \rhd S_1, T_1|T_2)$ and $puts(\Gamma c, \Pi_2 \rhd S_2, T_3|T_4)$ are well defined, then $puts(\Gamma c, \Pi_1 \rhd S_1, T_1|T_2) \precsim_{rbc}^{p} puts(\Gamma c, \Pi_2 \rhd S_2, T_3|T_4)$.

Now we shall give definitions for definability of external actions in GR. An external action is said to be definable if there is an appropriate test which can verify that such action is being performed.

Definition 14. Definability of input actions
We say that the input action α, which is of the form $R.n.m.c?(v)$ for any R, n, m, c *and* v is definable if for every finite set of names N and for every pair of channel names $SUCC$ and $FAIL$ there is a system term T_α $(N, SUCC, FAIL, \alpha)$ with the property that for every well-formed configuration $\Gamma c, \Pi \rhd S$ and $puts(\Gamma c, \Pi \rhd S, T_\alpha)$ is well-formed such that $fn(S) \subseteq N$, then

1. If $\Gamma c, \Pi \rhd S \rightarrow_a^{k,p} \Gamma c, \Pi \rhd S'$, then $puts(\Gamma c, \Pi \rhd S, T_\alpha) \Rightarrow^{k,p} \Gamma c, \Pi \rhd))[R]_{\boxtimes}^{0,1}\{m, m, nil @ SUCC\}|S'$.
2. If $puts(\Gamma c, \Pi \rhd S, T_\alpha) \Rightarrow^{k,p} \Gamma c, \Pi \rhd T$ that $\Gamma c, \Pi \rhd T \Downarrow^{0,p} [R]_{\boxtimes}\{m,m,nil @ SUCC\}$, $\Gamma c, \Pi \rhd T \not\Downarrow^{0,1}[R]_{\boxtimes}\{m, m, nil @ FAIL\}$, then $T \equiv [R]_{\boxtimes}^{0,1}\{m, m, nil @ SUCC\}|S'$ for some S' such that $\Gamma c, \Pi \rhd S \rightarrow_a^{\alpha\ k,p} \Gamma c, \Pi \rhd S'$.

Lemma 14 Every input action is definable.

Now we will give formal definition of definability of output actions similar to definability of input actions. We will then show that output actions are, in fact, also definable in GR.

Definition 15. Definability of output actions We say that the output action α, which is of the form $[R]_{\boxtimes}\{n,m,v @ c\}| T$ such that $\langle R^\Pi \rangle (m) = \{R\}$, is definable if for every finite set of names N and for every pair of channel names $SUCC$ and $FAIL$ there is a system term T_α $(N, SUCC, FAIL, \alpha)$ with the property that for every well-formed configuration $\Gamma c, \Pi \rhd S$ and $puts(\Gamma c, \Pi \rhd S, T_\alpha)$ is well formed such that $fn(S) \subseteq N$, then

1. 1. If $\Gamma c, \Pi \rhd S \rightarrow_a^{\alpha\ k,p} \Gamma c, \Pi \rhd S'$, then $puts(\Gamma c, \Pi \rhd S, T_\alpha) \Rightarrow^{k,p} \Gamma c, \Pi \rhd (new\ bn(\alpha))[R]_{\boxtimes}^{0,1}\{m, m, bn(\alpha) @ SUCC\}|S'$.
2. 2. If $puts(\Gamma c, \Pi \rhd S, T_\alpha) \Rightarrow^{k,p} \Gamma c, \Pi \rhd T$ that $\Gamma c, \Pi \rhd T \Downarrow^{0,1} [R]_{\boxtimes}^{0,1}\{m, m, bn(\alpha)@ SUCC\}$, $\Gamma c, \Pi \rhd T \not\Downarrow^{0,1}[R]_{\boxtimes}^{0,1}\{m, m, nil @ FAIL\}$, then $T \equiv_a^{\alpha\ k,p} (new\ bn(\alpha))[R]_{\boxtimes}^{0,1}\{m, m, bn(\alpha)@SUCC\}|S'$ for some S' such that $\Gamma c, \Pi \rhd S \rightarrow_a^{} \Gamma c, \Pi \rhd S'$.

Lemma 15 Every output action is definable.

Theorem 2 Every external action is definable.

Lemma 16 σ: $CN \to CN$, where CN is the set of channel names and if $\forall c, d \in CN$ such that $\sigma(c) = d$, then

1. $\Gamma c, \Pi_1 \triangleright S \to^{k,p} \Gamma c, \Pi_2 \triangleright T$ implies $\Gamma c, \Pi_1 \triangleright S\sigma \to^{k,p} \Gamma c, \Pi_2 \triangleright T\sigma$.
2. $\Gamma c, \Pi_1 \triangleright S \preccurlyeq_{rbc}^{p} \Gamma c, \Pi_2 \triangleright T$ implies $\Gamma c, \Pi_1 \triangleright S\sigma \preccurlyeq_{rbc}^{p} \Gamma c, \Pi_2 \triangleright T\sigma$ if the function σ is injective.

The next lemma enables us to recover the full use of the residuals obtained by using the definability tests to provoke actions, essentially stripping off the bound names inherited from output actions.

Lemma 17 Extrusion if $SUCC$ is fresh to both S and then $\Gamma c, \Pi_1 \triangleright$ *(new α)* $[R]_{\boxtimes}^{0,1} \{m, m, \alpha @ SUCC\} \mid S \preccurlyeq_{rbc}^{p} \Gamma c, \Pi_2 \triangleright$ *(new α)* $[R]_{\boxtimes}^{0,1} \{m, m, \alpha @ SUCC\} \mid T$ such that $\langle R^{\Pi 1} \rangle (m) = \{R\}$ and $\langle R^{\Pi 2} \rangle (m) = \{R\}$ implies $\Gamma c, \Pi_1 \triangleright S \preccurlyeq_{rbc}^{p} \Gamma c, \Pi_2 \triangleright T$.

Now we are ready to prove that our asynchronous efficiency pre-bisimulation completely characterises out touchstone equivalence.

Theorem 3. Full abstraction
$\Gamma c, \Pi_1 \triangleright S \preccurlyeq_{a}^{pe} \Gamma c, \Pi_2 \triangleright T$ iff implies $\Gamma c, \Pi_1 \triangleright S \preccurlyeq_{rbc}^{p} \Gamma c, \Pi_2 \triangleright T$

9   Conclusions

We justified our choice of asynchronous probabilistic efficiency pre-bisimulation for *GR* by proving that it coincides with the probabilistic reduction barbed pre-congruence. Finally, we proved that probabilistic reduction barbed pre-congruence can also be recovered from asynchronous probabilistic efficiency pre-bisimulation. It makes *GR* a full abstract language. We believe that such models help in prototyping the probabilistic routing algorithms/protocols in the large distributed networks and IoT's.

References

1. J. Woodcock, P.G. Larsen, J. Bicarregui, J. Fitzgerald, Formal Methods: Practice and Experience. ACM Comput. Surv. **41**(4), 19:1–19:36 (2009)
2. R. Milner, *Communicating and Mobile Systems: The π-Calculus* (Cambridge University Press, 1999)
3. K. Hofer-Schmitz, B. Stojanović, Towards formal verification of iot protocols: A review. Comput. Netw. **174**, 107233 (2020)
4. T. Margaria, B. Steffen, Leveraging Applications of Formal Methods, Verification and Validation. Distributed Systems: 8th International Symposium, ISoLA 2018, Limassol, Cyprus, November 5–9, 2018, Proceedings, Part III, vol. 11246. Springer (2018)

5. S. Ouchani, Ensuring the functional correctness of iot through formal modeling and verification, in *International Conference on Model and Data Engineering*, (Springer, 2018), pp. 401–417
6. U. Ozeer, Sala ün, G., Letondeur, L., Ottogalli, F.ç.G., Vincent, J.M., Verification of a failure management protocol for stateful iot applications, in *Formal Methods for Industrial Critical Systems*, ed. by M. H. ter Beek, D. Ničković, (Springer, Cham, 2020), pp. 272–287
7. S. Ouchani, Ensuring the functional correctness of IoT through formal modeling and verification, in *International Conference on Model and Data Engineering*, (Springer, 2018), pp. 401–417
8. K. Iwanicki, A distributed systems perspective on industrial IoT, in *2018 IEEE 38th International Conference on Distributed Computing Systems (ICDCS)*, (IEEE, 2018), pp. 1164–1170
9. P. Sethi, S.R. Sarangi, Internet of things: Architectures, Protocols, and Applications. J. Electr. Comput. Eng. **2017**
10. N. Sfeir, H. Sharifi: Internet of things solutions in smart cities
11. M. Stolpe, The internet of things: Opportunities and challenges for distributed data analysis. ACM SIGKDD Explor. Newslett. **18**(1), 15–34 (2016)
12. A.S. Tanenbaum, M. Van Steen, Distributed systems: Principles and Paradigms. Prentice-Hall (2007)
13. M. Gaur, R. Kant, A survey on process algebraic stochastic modelling of large distributed systems for its performance analysis, in *3rd International Conference on Eco-friendly Computing and Communication Systems (ICECCS)*, (2014), pp. 206–211. https://doi.org/10.1109/Eco-friendly.2014.49
14. Cerone, A.: Foundations of Ad Hoc Wireless Networks. Ph.D. thesis, School of Computer Science and Statistics, Trinity College Dublin (July 2012), https://www.doc.ic.ac.uk/acerone/works/thesis.pdf
15. G. Norman, D. Parker, *Quantitative Verification: Formal Guarantees for Timeliness, Reliability and Performance*, Tech. rep. (The London Mathematical Society and the Smith Institute, 2014)
16. Gaur, M.: A Routing Calculus: Towards Formalising the Cost of Computation in a Distributed Computer Network. Phd, Informatics, University of Sussex, U.K. (December 2008)
17. M. Gaur, A routing calculus for distributed computing, in *Proceedings of Doctoral Symposium Held in Conjunction with Formal Methods 2008*, ed. by E. Troubitsyna, vol. 48, (Turku Centre for Computer Science General Publication, 2008), pp. 23–32
18. M. Hennessy, *A distributed Pi-Calculus* (Cambridge University Press, 2007)
19. D. Sangiorgi, *An Introduction to Bisimulation and Coinduction* (Cambridge University Press, Cambridge/New York, 2012)
20. D. Sangiorgi, On the origins of bisimulation and coinduction. ACM Trans. Program. Lang. Syst. **31**(4), 15:1–15:41 (2009). https://doi.org/10.1145/1516507.1516510
21. M. Gaur, M. Hennessy, Counting the cost in the picalculus (extended abstract). Electronic Notes in Theoretical Computer Science (ENTCS) pp. 229:117–129 (2009)
22. K.S. Trivedi, *Probability and Statistics with Reliability, Queuing and Computer Science Applications*, 2nd edn. (Wiley, Chichester, 2002)

Steganography: Camouflaging Sensitive and Vulnerable Data

Srilekha Mukherjee and Goutam Sanyal

1 Introduction

The dynamically updating technologies feed on information that is often very sensitive and crucial. Therefore, securing its contents becomes a primary concern. Most often, the sensitive information acts as a target for the third-party untrusted sources. Such sources belong to groups of hackers and crackers. Though data hiding is a quite an old area of research, rigorous exploration are producing new strategies for the sole purpose. A number of researchers are coming up with unique methodologies of data hiding. Cryptography [1], watermarking, and steganography are the highlighted areas in the field of security. Each of the stated domains has various methodologies that are beneficial in procuring the data hiding procedures. Cryptography is a very efficient and remarkable domain that features the encryption of data/information. It protects the content of the data by making it absolutely unreadable. Therefore, for any third-party sources, the information appears to be illegible and meaningless. The policy of watermarking [2] is to embed watermarks or authorized content or mark within the source so that it doesn't lose its original authorship under any circumstances. Even if the data is tampered, the watermark stays intact within and this helps to identify the legal owner of any such document.

While the other two stated areas target maintaining security in one aspect, steganography [3] targets the security concept by completely camouflaging the actual content of data that is hidden. This actually conceals its very existence so no third-party realizes whether something is hidden or not. Steganography has its

S. Mukherjee (✉)
Department of Computer Science and Engineering, University of Engineering and Management, Kolkata, India

G. Sanyal
Department of Computer Science and Engineering, National Institute of Technology Durgapur, Durgapur, India

93

applications ranging through a wide area that prioritizes covert data communication. There are certain terminologies related to steganography which helps to categorize and classify its types. The cover/carrier object is one such term that refers to the outer medium that carrying information. It appears to be seemingly unimportant to the outsiders. The stego object is another terminology, which refers to the cover data with the secret information inserted within it, or rather this refers to the file carrying the sensitive information that is transmitted through the public channel.

Based on the cover medium, steganography is categorized into the following types (shown in Fig. 1):

- Text steganography: Here, the cover medium within which the secret data is to be hidden is text medium. The content of the text, in text steganography is exploited so as to hide data.
- Image steganography: The carrier medium in this case is an image and data is hidden with it. The features of image are the targeted concern in this case, for the purpose of data hiding.
- Audio steganography: According to the other cases, the carrier file here is an audio file. Data is hidden within the audio file; thus the name being audio steganography.
- Video steganography: A video file serves to be a carrier here. Data is hidden within the selected seemingly unimportant video files. The frames of the video are the prime target in the case of video steganography.
- Protocol steganography: The secret data bits are supposedly hidden within the unused network bits, especially within used headers of transmitting protocols. Henceforth, the name derived as protocol steganography.

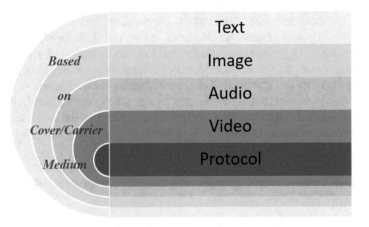

Fig. 1 Types of steganography (on the basis of cover/carrier medium)

1.1   Beginning of the Era of Steganography:
The Prisoners Problem

In 1983, the scientific and practical study of steganography had actually begun with reference to the open literature. The famous reference to the prior stated study of steganography is Simmons Prisoners' Problem. This study was formulated at a point when Simmons described this problem in terms of some kind of communication being taking place in a prison. To demonstrate further, consider two prisoners, namely, Alice and Bob, are grounded in prison. They are trying to hatch a plan to escape via some communicating channel that is public. Moreover, all the information exchanges that occur between them are always under strict surveillance of their warden, named as Eve. Eve may put these prisoners into solitary confinement, even upon suspicion of some kind of manipulated communication. Also, the two prisoners are totally aware of such reality. Henceforth, before going into custody, they had already shared some secret key code word. This will be used by them for now going to venture and add a hidden meaning to their seemingly innocent messages. Also, they have to make sure that Eve will not be able to decipher their ideas without the secret key codeword. Both Alice and Bob achieve success if they manage to communicate only by exchanging some information that allows them to execute their plan to flee. Also this should take place without Eve being skeptical in any way. This is done via the use of steganography. This problem can also be related to any two security officials who are attempting a secret communication through any public channel, which may be lined with the occurrence of a dangerous spy.

The above-stated illustration of the strategy is pictorially depicted next in Fig. 2. This may be considered to be the supposed base framework of the Prisoner's Problem. Alice successfully embeds a secret message (say, 'm') into the chosen carrier or cover object (say, 'c'). This hence gives the stego-object (say, 's'). The purposefully generated stego-object('s') is then sent through a public medium or communication channel to reach the other person Bob. In any sort of pure steganographic framework, this embedding technique particularly for the secret messages is unknown to Eve. It may only be shared as a secret code between the two people, Bob and Alice. On the other hand, in case of some private key steganography, the sender and receiver, i.e. Alice and Bob, share a secret key. Such key was used for the purpose of embedding the private message. Now, this secret key, as an example, may be any sort of password that procures some pseudo-random number generator for the purpose of selecting pixel locations in say some image cover-object for carrying out the embedding of the message. The warden, Eve, is completely unaware of such secret keys that are already shared between Bob and Alice share. But she is said to be aware of the specific algorithm or technique that they might be using for embedding any kind of secret message. Basically, in case of public key steganography, both Bob and Alice utilize and use pairs of private-public key, henceforth knowing their individual default public key.

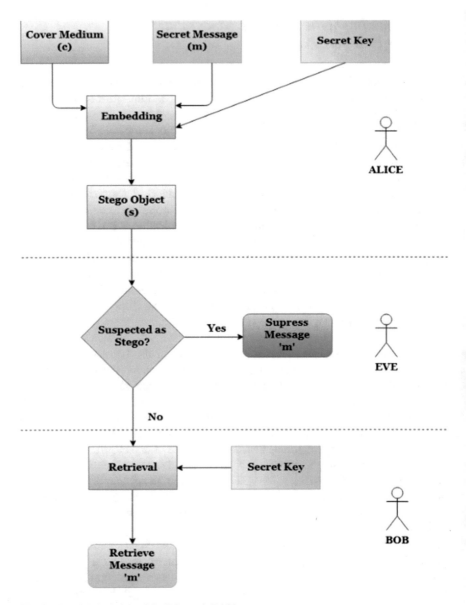

Fig. 2 Pictorial depiction of the Prisoner's Problem

1.2   A Generalized Framework of Steganography

Let's study a basic structure of the working for any general steganographic system. A flow diagram is shown below.

Figure 3 represents the basic model for working of any general steganographic system. On the sender side, the sender sends the secret object to the respective recipient through a public communication medium. This medium or channel is considered to be insecure. The sender, on his/her side, embeds the secret object into the respective cover object by using some kind of embedding strategies or method. This step produces the stego object, which is actually sent. The key is supposedly optional. Henceforth, it may or may not be used for the purpose of embedding. The stego object is finally sent to the receiver. The receiver, on his/her side, uses the respective relevant extraction algorithm to retrieve the secret object from the stego received..

1.3   Categorization in Steganography

Steganography can also be generally categorized into three categories:

- Pure steganography: It is the type where there is no stego-key. This is based on a simple assumption that no unwanted third-party source is aware of the communication being taking place.
- Secret key steganography: This is the type where the stego-key is exchanged beforehand, i.e. this exchange takes place prior to any communication taking place. Also, this type is very much susceptible to some kind of interception.
- Public key steganography: In this kind, both public and private keys are used so as to achieve a secure communication.

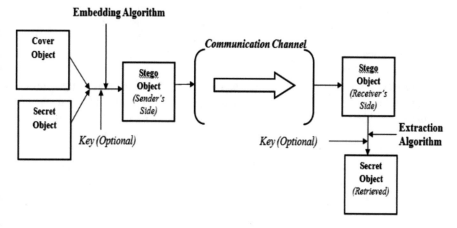

Fig. 3 Working model for a generalized steganographic system

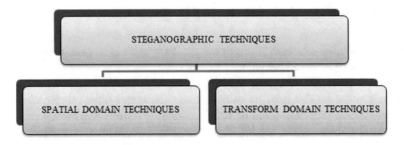

Fig. 4 Types of steganographic techniques

Also by the way in which the secret bits are embedded within the cover medium, we may also classify steganographic techniques into several types, namely, spatial and transform domain (as per Fig. 4). These are discussed in detail in the section of case studies (Sect. 2).

1.4 Criteria for a Good Steganographic Approach

The primary criteria of a good steganographic approach are classified and grouped into the stated three parameters. They are payload or embedding capacity, robustness, and security.

- **Embedding capacity (payload):** It indicates the net or total volume of any specific secret message that is supposed to be hidden within the carrier/cover to ensure a secured communication. Therefore, the maximum amount of the hidden data that is transmitted is its said capacity [4]. This must go hand in hand with minimization of perception of the same secret hidden message within that of the stego, i.e. without affecting the quality.
- **Robustness:** This is known to be the ability of the formed steganographic systems to resist the attacks. Such attacks may be made on the stego object. Henceforth, it is actually necessary to develop the steganographic methodology in such a way that it is competent of combating the conventional manipulations those may appear.
- **Security:** By security of any of the steganographic system, it actually signifies the inability of any such attacker to detect [5] the embedded/hidden data within the stego object. Any sort of distortion in quality of the carrier may certainly raise some suspicions of eavesdroppers. Therefore, in order to avoid the further detailed screening as well as detection, the contents of that hidden information must be kept seamless or invisible. Maintaining imperceptibility should be the target concern here. Imperceptibility signifies high resemblance between the cover and stego object.

2   Case Studies in Literature

2.1   *Spatial Domain Techniques*

The spatial domain techniques are those techniques that exploit and modify the resident bits in the spatial domain. For the sole purpose of facilitation of data hiding, they utilize or rather changes the resident pixels in the cover/carrier file. From the perspective of embedding/extraction complexity, the method that comes under this specific domain is some of the few simplest techniques. In the image domain, data hiding by the method of least significant bit (LSB) [6] is one of the traditional methods, which is well known to hide data by directly replacing the LSB level bits with those of the secret bits. The LSBs always have least or minimum weightage, which actually contributes to the net value of pixel intensity. Hence any change made to such LSB fields will not contribute much to the actual change in intensity. This method acts as a base method for many other approaches in this domain. When a slight change is made in the RGB pixel values, it results in the production of another new variant of a colour. This new formed one is so close to its base colour that their supposed resultant differences are simply next to impossible to visualize. Moreover, such tactics are mostly applicable to the compression schemes of images that are mainly lossless, like that of an image having TIFF format extensions, etc. In case of those schemes that are supposedly of lossy compression [7], like may be that of JPEG, etc., a very few bits of the message might get lost while the phase of compression is ongoing. This approach on one side might supports high embedding capacity, but it is also quite affected by certain characteristic changes such as cropping or compression of chosen test images.

There are few extensions to this method. One such method is proposed by Zhang et al. [8]. This is primarily one method for LSB matching (LSBM). As goes the name, this method slightly induces few changes that appends to the main or original scheme. It suggested that this new embedding scheme was based on a simple logic of plus/minus. The secret stream of the data bits is matched with that of the LSB of its respective next resident carrier or cover pixel. This is done based on some pseudo-random order. When an existent match is found, no change is supposedly made. Else the cover or carrier pixel value, at a random order, is incremented or decremented by 1. Therefore, the natures of the bit values of the secret as well as carrier or cover are the prime determining factors. There is another LSB-based technique, which uses truth table and determinate array RGB indicators for the pixel manipulation. Here, we may say that the least significant bits of any of the single channel may also indicate presence of data in the other channels. This sort of technique promotes high embedding capacity. In the work of Weiqi Luo et al. [9], the authors describe a scheme of edge adaptivity which can further select the specific embedding regions. This is done in accordance with the size of the secret message along with the disparity between two selected consecutive pixels in the carrier or cover image. While embedding data, this scheme initializes some parameters at first. Such are used mainly for estimating the net capacity of the individually selected regions. At the end, the stego image is obtained only after the stage of pre-processing. Here, there is an application of region adaptive scheme to the bits of spatial LSB domain. Luo et al. had basically proposed a new steganographic method, which is edge adaptive (EA-LSBMR). In this methodology, the carrier is pre-processed in the initial stage. Thereafter, some specific embedding regions are selected in a kind of adaptive way. All these are decided as well as resolved by some pseudo random number generator (PRNG). Again, the lengths of hidden message as well as that of the pixel-pairs differences are both taken into consideration. Moreover, the process of data hiding here is performed on the basis of the strategy of least significant bit matching revisited (LSBMR). The triple-A algorithm [10] may also be stated as one other efficient methodology to channelize data hiding. All the concepts that are used here are similar to that used in the LSB approach. The one and only difference that lies here is in the fact that, here, more and better degree randomization is facilitated. Also, the selection of the net or total number of bits as well as colour channels that are to be used is more randomized. This is also done to increase the overall capacity along with the security of the system. Banharnsakun [11] has introduced one very efficient method using concepts of the artificial bee colony (i.e. ABC) methodology. This is done in order to enhance the pros of LSB-based steganographic approach. Here, for the purpose of optimizing the assignment of blocks mainly for channelizing the hiding procedure, this approach of ABC is applied.

Wu et al. [12] proffered an approach, which takes into consideration some kind of synthesis of reversible texture. Such operation of the stated texture synthesis is blended within the steganography in order to embed the secret data. Instead of using any specific carrier or cover, this particular algorithm masks and blends the source texture image, thus embedding several messages via the aforesaid synthesis. Mielikainen [13] has proposed another different modification made to that of the least significant bit matching (LSBM) strategy, named as least significant bit matching revisited (LSBMR). Such strategy utilizes some choice to define a binary function for that of the two carrier pixels to a nearly suitable value. This suitable value of the stated set binary function actually carries the information. Such a modified method certainly promotes same level of payload embedding as that of the LSB matching one. The only with this is a lesser modification might be made to the carrier cover. Another variant of edge adaptive steganographic technique that supports quite high payload is given by Chakraborty et al. [14]. In this, a selective kind of area from the chosen carrier has to be anticipated with the immediate aid of a modified median edge detector or MMED predictor. This is done so as to embed the private data in its binary form.

The method of pixel value differencing (PVD) [15] is another mentionable efficient method for the purpose of data hiding. Such a method primarily considers each of the adjacent pixel difference for the purpose of hiding bits. In this stated approach, the carrier is framed into distinct non-overlapping subsections that embody two of the connecting pixels. Later, the pixel difference between each of the pairs of the extant blocks has to be altered in order to ingrain data bits. In its extraction phase, the received stego is partitioned with the help of the same stated method, with the aid of the formed range table. In general, this specific procedure deals with adding or camouflaging large number of data bits within that of image regions of high variance. For example, we may say that the areas near 'the edge' could be considered using the difference of values between that of the resident neighbouring pixels. Again, this approach may be improved further when clubbed with another least significant bit embedding. Now, having the PVD method acting as a pivot, there are various other proposed methodologies that are based on PVD [16]. Among them is the one given by Chang et al. [17]. There is a unique three way of considering the pixel value differencing in this approach. Such strategy is considered as a marked improvement over the base or traditional PVD approach. Here, with respect to the generated values of embedding capacity as well as peak signal-to-noise ratio (PSNR), such a technique is proved to be adept. Another method for hiding data is by using the gray level modification (GLM) method. This was proposed by Potder et al. in literature. Such a technique proceeds by reshaping or remodeling each gray level of the resident pixel values of carrier. This is done so as to map (not embed or hide) the individual data bits. For the purpose of mapping data, this basically uses the simple concept of even/odd numbers. We may say proceed by considering 0's for the case of even pixels and 1's for the respective odd values. The described method hinges on to some classic mathematical function. Also a batch of pixels is chosen and their respective gray level values are collated to the bit stream of the sensitive message that is supposed to be mapped within the carrier. There is a strategy of merging the PVD strategy with the GLM technique [18]. This is done mainly to dilate and improve the embedding capacity. Such methodology is proposed by Safarpour et al. [19]. It generally uses a hybrid way to combine and merge both the concepts of PVD along with GLM. Here, in the first phase a group of data is hidden with the aid of the technique of PVD. Again two types of pixel difference from the adjacent pixel pairs are taken into consideration here. One is the case where both the pixel pairs are known to be as even and the second case is where both are known as odd. This stage is followed by hiding the remaining data with that of the procedure of GLM. Also, it is made sure that during this ongoing second phase, no data that was hidden in the first phase gets lost. The final generated capacity of the hidden data bits is found to be quite high in comparison to each of the individually known processes.

A new fast way of matrix embedding technique has been defined by Chao Wang et al. [20]. This is done mainly by adding and appending some of the referential columns of each respective parity check matrix. Finally, the net estimated computational complexity of the embedding procedure may be significantly reduced with the help of this method. Such an advantage also helps to achieve a high embedding efficiency as well. Also, the embedding speed is very high. One of the methodologies by Ahmad et al. [21] channelizes the hiding within the spatial domain of the gray scale carrier. Here, in the initial stages, the carrier is fragmented into a number of identical block sizes. Depending upon the concept of the prevailing number of 1's in the left four bits of the resident pixels, the chosen private message is hidden. Also, data bits are hidden only in the outskirts of the prior obtained uniform segments. The results of this stated approach give better values in comparison to many of the traditional methods. The pixel mapping method (PMM), as described by Bhattacharya et al. [22] chooses and elects the embedding pixels by virtue of some mathematical function. This mainly depends on definite intensity values of such pixels. Also, in accordance with the stated concept, its respective eight neighbours are all elected from a direction, which is logically counter clockwise. Now, before embedding begins, a checking is made in rigorous order so as to find out whether any of such randomly chosen pixels or any of its adjacent neighbours does lie in the image boundary. Thereafter, by taking into consideration each individual intensity values as well as the number of binary 1's prevailing in the pixel, the hiding of data is facilitated. This is proceeded by mapping of two bit of secret data or information within the individual marked neighbour pixels. Such technique also supports high embedding capacity as in few cases it might also variate the net number of individual bits to be hidden.

Data hiding approaches that use cryptography along with numerous tricks of steganography are also proposed in literature. One such method uses AES cryptography [23]. It generally describes a two layer concept for the purpose of facilitating hiding of sensitive data or information. This increases its hiding capacity on the whole. Also the overall security is enhanced. Yu C. et al. [24] have proposed one excellent technique of facilitating four-image encryption. Here, the image encryption scheme is proposed with the aid of several factors. These are the quaternion Fresnel transform [25] along with some computer-generated hologram as well as a logistic-adjusted-sine map (LASM) of two dimensions. In such technique, four images, which are symbolized by quaternion algebra, are channelized and processed in a vector approach by the help of quaternion Fresnel transform. Hereafter the amplitude, which was actually constructed by the stated QFST-transformed image components, is overall encoded by Fresnel transform along with the two random phase masks (RPM). These two are also virtually independent.

2.2 Transform Domain Techniques

In the frequency or transform domain, the selected carriers are first transformed with the aid of any of the frequency-oriented mechanisms. Such mechanisms may include discrete wavelet transform (DWT) [26], discrete Fourier transform (DFT) [27], discrete cosine transform (DCT) [28], or some other similar transform operation which might transform the spatial values into the coefficients of the transform/ frequency domain. The next stages are the ones that facilitate the embedding of the secret data or messages within that of the coefficients in frequency form.

For any of the chosen medium, a steganographic algorithm that might make fewer changes while embedding data bits always appends less amount of additional noise [29]. Henceforth, the secret data will be less detectable. This of course may be said in relative to some algorithm that might introduce or add higher additive noise by promoting more amounts of changes. Therefore, following the above concept, Crandall introduced the use of 'matrix encoding' [30] or error control coding technique. Here, we supposedly consider 'q' as bits of message are to be hidden within a group comprising '$2^q - 1$' pixels of carrier data. Again, per group, some kind of noise, which lies in the range of '$1-2^{-q}$' may be added on an average scale. The highest embedding capacity or payload, which may be attained, is '$\dfrac{q}{2^q - 1}$'.

Taking an example, let's consider 2 bits of some secret data or message that has to be hidden or embedded within some formed group of 3 pixels that simultaneously adds and appends some noise to each group, say, 0.75 on an average scale. In this case, the highest hiding capacity that might be reached is 2/3 = 0.67 bits/pixel. Probably one of the most prominent applications of matrix encoding is said to be the F5 algorithm [31]. It uses the matrix encoding technique in order to decrease the value of embedded noise. For the exclusive purpose of masking the secret data bits, F5 algorithm does uses the coefficients of discrete cosine transform from the distinctly chosen carrier. It always masks the data with rounding off the existing coefficients to their most adjacent data bits. Apparently, without replacing the least significant bits of any of those quantized coefficients of DCT with that of the secret data bits, it actually drops this absolute face value of those coefficients by 1. Also this leaves a provision of some case when necessary to be readjusted. During the overall process of embedding, the net number of any of the AC coefficients that is non-zero and the lengths of messages do influence the best kind of matrix embedding. Also, this reduces the total alterations on the carrier/cover.

Jpeg–Jsteg [32] is very popularly known to be a data hiding tool. It is primarily based on JPEG images. In this technique, the secret bits are masked or hidden in those low-frequency subsections of the quantized DCT coefficients [33], which have no values as 0, 1, or −1. Also, the specific sequence of embedding, which is supposedly followed here, is the order of some zigzag scan. Primarily, the JPEG images have the compression techniques as their key advantage over and above several other formats. Also by using such schemes of compression, images with high colour quality could be very easily stored within several small files. Moreover, embedding into the various DCT coefficients does not cause any noticeable or rather visible effects or any other detectable statistical changes [34]. By facilitating any embedding in the transform domain, it may not always point towards achieving those transform coefficients over any block (of some size, say, 8 × 8) as it might often be done in techniques for compression of JPEG. Such techniques designed may always consider the transforms on the image as a whole. Several other algorithms related to block-based JPEG domain as well as strategies of wavelet-based embedding are also stated and defined in literature.

Yet Another Steganographic Scheme or YASS [35], which does belong to the JPEG steganography scheme, does not insert secret data bits directly in those DCT coefficients. Here, the input spatial representation is broken down into several blocks of fixed sizes. These are termed as big blocks or B-blocks. Within each of such blocks, various sub-blocks of 8×8 sizes (called as embedding host blocks or H-blocks) are selected with the aid of some secret key for the purpose of applying DCT. The secret data are formerly encoded with the help of specific error correction codes. These are later embedded within those of the obtained DCT coefficients of that of the H-blocks with the aid of Quantization Index Modulation or QIM [36]. Further, on applying inverse DCT to these obtained H-blocks, the image is compressed in a form of JPEG image. Later, during data extraction stage, the stegos are first JPEG-decompressed into their spatial domains. Thereafter, the hidden data is finally extracted from that of the concerned DCT coefficients of the above obtained H-blocks. As the position of obtained H-blocks does not overlap with that of the 8x8 grids of the JPEG, those embedding artefacts, which might have been caused by the application of the approach of YASS, are not reflected in these DCT coefficients of JPEG. OutGuess [37] is another efficient technique in this domain. It also has its embedding process branched into some of the phases. In the beginning, it proceeds by embedding the private data bits within the least significant bits of the quantized DCT coefficients. This is done by following a random walk alongside with the skipping of the 1's and 0's. Next, during the onset of post embedding, several corrections are made and issued to the obtained coefficients. These were not chosen during the ongoing stage of embedding. Also, this is done to match the DCT histograms of the individual stegos with their respective covers.

In another approach a secure as well as robust steganographic algorithm with video [38] as medium is proposed in discrete cosine transform (DCT) in conjunction with discrete wavelet transform (DWT) domains. This is found to be built on multiple object tracking or MOT algorithm along with error correcting codes. For the purpose of encoding, the secret data is initially pre-processed by application of both Hamming [39] and BCH codes (i.e. Bose, Chaudhuri and Hocquenghem). At first, MOT algorithm (motion based) is applied on carrier videos so as to identify the prime regions of interest or ROI within the actual moving objects of the video. Next, the data embedding is channelized by masking the private bits within the DCT and DWT coefficients of the previously obtained motion regions (based on foreground masks). Another method, as proposed by Sheisi et al. [40], embeds data within the obtained DCT coefficients pertaining to the middle frequencies. In the middle frequency area, it replaces the coefficients of the LSB with secret data bits starting from the very last coefficient. Moreover, the data hiding capacity obtained from this method is found to be less than that of the JSteg algorithm. A new transform domain steganographic methodology [41] for digital images is proposed based on integer wavelet transform or IWT. This technique uses a chaotic map, which may be said to be a modified form of logistic map that increases individual key length along with the security of the proposed approach. Experimental results of this method show that it has high potential of masking data within any chosen image that may be used as the cover/carrier media. One novel approach based on the colour image steganography is proposed by Rabie et al. [42]. This method takes into inspection the existing homogeneity of the spatial distribution within the carrier. This proposed approach observes that each space, which is reserved for the purpose of embedding bits of the confidential message, fluctuates with the change in statistical properties of each of the selected carrier. It divides the cover into some segments, which may adapt in accordance with the existing correlation of that structure of image. This forms blocks of carrier image (of variable sizes) that are statistically stationary. It also exploits the embedding scheme of quad-tree adaptive region or QTAR to demonstrate the stated statistically stationary segments of cover with respect to obtained correlation of pixels of the image. A moderately optimized capacity of embedding is thus ensured in comparison to the methods with fixed block size sectioning of covers.

3 Conclusion

There are a huge variety of steganographic approaches in literature. Also, many recent researches are coming up in this domain. Gradually the huge void that existed while protecting sensitive information is reducing day by day. More and more newer techniques are coming up to bridge such gaps. In this paper, we have summed up some works done in this area of research. They may have multiple applications in the field of data hiding. Also, they might help to identify and bring out other new ideas and also improve that of the old ones.

References

1. J. Ding, J.-P. Tillich (eds.), *Post-Quantum Cryptography – 11th International Conference, PQCrypto 2020, Paris, France, Proceedings*, Lecture notes in Computer Science, vol 12100 (Springer, 2020)
2. S. Kumar, B.K. Singh, M.A. Yadav, Recent survey on multimedia and database watermarking. Multimed. Tools Appl. **79**, 20149–20197 (2020)
3. M. Sharifzadeh, M. Aloraini, D. Schonfeld, Adaptive batch size image merging steganography and quantized Gaussian image steganography. IEEE Transactions on Information Forensics and Security **15**, 867–879 (2020)
4. M.C. Kasapbaşi, W. Elmasry, New LSB-based colour image steganography method to enhance the efficiency in payload capacity, security and integrity check. Sādhanā **43**, 68 (2018)
5. V. Sedighi, R. Cogranne, J. Fridrich, Content-adaptive steganography by minimizing statistical detectability. IEEE Trans Inform Forens Secur **11**(2), 221–234 (2016)
6. S.T. Veena, S. Arivazhagan, Quantitative steganalysis of spatial LSB based stego images using reduced instances and features. Pattern Recogn. Lett. **105**, 39–49 (2018)
7. N. Johnston, D. Vincent, D. Minnen, et al., Improved Lossy image compression with priming and spatially adaptive bit rates for recurrent networks, in *The IEEE conference on computer vision and pattern recognition (CVPR)*, (2018), pp. 4385–4393
8. W. Zhang, X. Zhang, S. Wang, A double layered plus-minus one data embedding scheme. Signal Processing Letters, IEEE **14**(11), 848–851 (2007)
9. W. Luo, F. Huang, J. Huang, Edge adaptive image steganography based on LSB matching revisited. IEEE Trans Inform Forens Secur **5**(2), 201–214 (2010)
10. A. Gutub, A. Al-Qahtani, A. Tabakh, Triple-A: Secure RGB image steganography based on randomization, in *IEEE/ACS international conference on computer systems and applications*, (Rabat, Morocco, 2009), pp. 400–403
11. A. Banharnsakun, Artificial bee colony approach for enhancing LSB based image steganography. Multimedia Tools Appl. **77**(20), 27491–27504 (2018)
12. K.C. Wu, C.M. Wang, Steganography using reversible texture synthesis. IEEE Trans. Image Process. **24**(1), 130–139 (2015)

13. J. Mielikainen, LSB matching revisited. Signal Process Lett, IEEE **13**(5), 285–287 (2006)
14. S. Chakraborty, A.S. Jalal, C. Bhatnagar, LSB based non blind predictive edge adaptive image steganography. Multimedia Tools Appl. **76**(6), 7973–7987 (2017)
15. Z. Li, Y. He, Steganography with pixel-value differencing and modulus function based on PSO. J. Inf. Sec. Appl. **43**, 47–52 (2018)
16. M.A. Hameed, S. Aly, M. Hassaballah, An efficient data hiding method based on adaptive directional pixel value differencing (ADPVD). Multimedia Tools Appl. **77**(12), 14705–14723 (2018)
17. Y.P. Lee, J.C. Lee, W.K. Chen, K.C. Chang, I.J. Su, C.P. Chang, High-payload image hiding with quality recovery using tri-way pixel-value differencing. Inf. Sci. **191**, 214–225 (2012)
18. K. Muhammad, J. Ahmad, H. Farman, Z. Jan, M. Sajjad, S.W. Baik, A secure method for color image steganography using gray-level modification and multi-level encryption. TIIS **9**(5), 1938–1962 (2015)
19. M. Safarpour, M. Charmi, Capacity enlargement of the PVD steganography method using the GLM technique. CoRR abs/1601.00299 (2016)
20. C. Wang, W. Zhang, J. Liu, N. Yu, Fast matrix embedding by matrix extending. IEEE Trans Inform Forens Secur **7**(1), 42–49 (2012)
21. T. Ahmad, M. Abdullah, A new approach for data hiding in gray-level images, in *SPPRA '08, Proceedings of the Fifth IASTED International Conference on Signal Processing, Pattern Recognition and Applications*, (2010), pp. 48–53
22. S. Bhattacharyya, G. Sanyal, Study and Analysis of Quality of Service in Different Image Based Steganography Using Pixel Mapping Method. Int. J. Appl. Inform. Syst. **2**, 42–57 (2012)
23. N.A. Al-Otaibi, A.A. Gutub, 2-Leyer security system for hiding sensitive text data on personal computers. Lecture Notes on Information Theory **2**(2), 151–157 (2014)
24. C. Yu, J. Li, X. Li, et al., Four-image encryption scheme based on quaternion Fresnel transform, Chaos and computer generated hologram. Multimed. Tools Appl. **77**(4), 4585–4608 (2017)
25. D. Kelly, Numerical calculation of the Fresnel transform. Journal of the Optical Society of America. A, Optics, image science, and vision **31**, 755–764 (2014)
26. M.A. Wakure, A.N. Holambe, A discrete wavelet transform: A steganographic method for transmitting images. Int. J. Comput. Appl. **129**(5) (2015)
27. H. Okada, K. Umeno, Randomness evaluation with the discrete Fourier transform test based on exact analysis of the reference distribution. IEEE Trans. Inform. Forens. Secur. **12**(5), 1218–1226 (2017)
28. X. Zhou, Y.Y. Bai, C. Wang, Image compression based on discrete cosine transform and multistage vector quantization. Int. J. Multimedia Ubiquit. Eng. **10**(6), 347–356 (2015)
29. Y. Li, Z. Li, K. Wei, W. Xiong, J. Yu, B. Qi, Noise estimation for image sensor based on local entropy and median absolute deviation. Sensors **19**(2), 339 (2019)
30. L. Liu, A. Wang, C.C. Chang, Z. Li, A secret image sharing with deep-steganography and two-stage authentication based on matrix encoding. I. J. Netw. Secur. **19**(3), 327–334 (2017)
31. T. Vaccon, K. Yokoyama, A tropical F5 algorithm, in *ISSAC*, ed. by M. A. Burr, C. K. Yap, M. S. El Din, (ACM, 2017), pp. 429–436
32. T.H. Thai, R. Cogranne, F. Retraint, Statistical model of quantized DCT coefficients: Application in the Steganalysis of Jsteg Algorithm. IEEE Trans. Image Process. **23**(5), 1980–1993 (2014)
33. P. Forczmański, W. Maleika, Predicting the number of DCT coefficients in the process of seabed data compression, in *Computer Analysis of Images and Patterns, CAIP 2015*, Lecture notes in computer science, ed. by G. Azzopardi, N. Petkov, vol. 9256, (Springer, Cham, 2015), pp. 77–87
34. M. Aktar, M.A. Mamun, M.A. Hossain, Statistical similarity based change detection for multitemporal remote sensing images. J. Electr. Comput. Eng. **3123967**, 1–8 (2017)
35. L. Yu, Y. Zhao, R. Ni, G. Cao, A channel selection rule for YASS. Sci. China Inf. Sci. **57**(8), 1–10 (2014)

36. Z. Kricha, A. Kricha, A. Sakly, Accommodative extractor for QIM-based watermarking schemes. IET Image Process. **13**(1), 89–97 (2019)
37. T.H. Thai, Optimal detection of outguess using an accurate model of DCT coefficients, in *WIFS*, (IEEE, 2014), pp. 179–184
38. R.J. Mstafa, K.M. Elleithy, E. Abdelfattah, A robust and secure video steganography method in DWT-DCT domains based on multiple object tracking and ECC. IEEE Access **5**, 5354–5365 (2017)
39. F. Li, A class of cyclotomic linear codes and their generalized Hamming weights. AAECC **29**(6), 501–511 (2018)
40. H. Sheisi, J. Mesgarian, M. Rahmani, Steganography: Dct coefficient replacement method and compare with JSteg algorithm. Int. J. Comput. Electr. Eng. **4**(4) (2012)
41. M.Y. Valandar, P. Ayubi, M.J. Barani, A new transform domain steganography based on modified logistic chaotic map for color images. J. Inf. Sec. Appl. **34**, 142–151 (2017)
42. T. Rabie, I. Kamel, Toward optimal embedding capacity for transform domain steganography: A quad-tree adaptive-region approach. Multimedia Tools Appl. **76**(6), 8627–8650 (2017)

Technologies for the Rehabilitation of People Affected with CVM: A State-of-the-Art Report

Arpita Ray Sarkar, Goutam Sanyal, and Somajyoti Majumder

1 Introduction

Cerebrovascular malfunction (CVM) refers to the disorders related to the blood vessels (arteries and veins) that supply blood to the brain, and as a result of these disorders, the blood supply to the brain is interrupted. A few such common CVMs include stroke, transient ischemic attack, carotid stenosis, vertebral stenosis, intracranial stenosis, aneurysms and vascular malformations. CVM is reported as the third cause of death worldwide [1]. Apart from death, CVM causes motor disabilities, loss of cognitive functions, communication difficulties, memory loss, etc. People affected by CVM require prolonged care under constant supervision and assistance. Immediately after CVM, these people are treated through medication, and after that necessary rehabilitation process is initiated in the hospital. Rehabilitation is a continuous process that aims to maximise the independence of a person through the improvement of health conditions and quality of life by minimising the consequences of the disease or disorders or malfunctions [2, 3]. Rehabilitation includes rehabilitation nursing, physical and occupational therapy, speech–language pathology, audio recreation therapy, nutritional care, rehabilitation counselling and social work [4, 5]. Services from professionals are not only required at the hospitals but also at home after release from hospitals. Both human and technological resources are required for rehabilitation. The aim of rehabilitation is independent living of a person affected by CVM, but such independence cannot be achieved without interdependence. Dependence on the physiotherapists, occupational therapists and the caregivers becomes essential during the earlier stage after the incident due to the

A. R. Sarkar (✉) · G. Sanyal
National Institute of Technology, Durgapur, West Bengal, India

S. Majumder
CSIR – Central Mechanical Engineering Research Institute, Durgapur, West Bengal, India

109

present policy of earlier discharge from the hospital and treatment at home. But, the support and assistance from the physical and occupational therapists can only be availed during the period of hospitalisation or a few days after release from the hospital due to the lack of sufficient numbers of such health professionals. It has been reported that there is a huge gap between the supply and demand of health workers or professionals (other than physicians) [6]. The density of health workforce for the countries with lower incomes (which include sub-Saharan and central African countries and Asian countries including Bangladesh, Nepal, India and eastern Mediterranean countries) is only 0.5 workers per 10,000 habitants. Figure 1 clearly shows the critical shortage of health professionals in developing countries. India is also suffering from such shortage.

As per the report by the Word Health Organization (WHO), the number of physiotherapists in the African, South-East Asia and Eastern Mediterranean regions is well below 30 per one million population [8]. Simultaneously, the data from 71 countries worldwide reveal the acute shortage of (registered) occupational therapists, even in the higher income countries, whereas the recommended minimum number of occupational therapists per one million population should be 750 occupational therapists by the World Federation of Occupational Therapists (WFOT). Besides, due to a limited number of such professionals in the majority of the countries (except few developed countries) and longer duration of the recovery period, it may not be possible to engage individual professional or specialist for the individual patient. In such cases, initially, the affected persons are treated and assisted through such professionals and specialists. Later, with passing of time and achieving a level of recovery, affected persons may depend mostly on different technologies that support and assist rehabilitation.

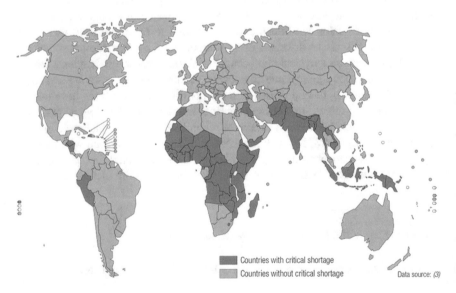

Fig. 1 Worldwide shortage of health professionals. [7]

Technologies are playing a major role in restoring or improving the affected function of the person undergone CVM or other similar diseases or injuries. Limited or lack of mobility of limbs or body and difficulties in communication due to the loss of memory, language problem, affected thought process, etc. are the most common after-effects of CVM. Computer hardware and software; electronic devices and systems; mechanical devices and aids; and, above all, material science have enriched the field of rehabilitation. Broadly, these technologies can be classified into two distinct categories: rehabilitative and assistive technologies. The rehabilitative technologies aim to restore or recover or improve the lost or affected function of the limbs/body parts or body due to certain disabilities. The assistive technologies may be small aids, devices or enhancements that help the people with disabilities to perform certain activities or tasks at homes, hospitals, schools or workplaces. Such technological resources, developed so far, have also been presented.

2 Rehabilitation for People Affected by CVM

Rehabilitation refers to an approach to regain skills, abilities and knowledge that have been lost due to acquired disabilities and to reach maximum self-independence. It involves several interconnected steps, as shown in Fig. 2, and known as rehabilitation cycle [9]. It starts with the identification of the problems and needs and then relates the problems to relevant factors of the individual and environment. After those, proper goals are defined for achievement through appropriate therapies. Accordingly, interventions are planned and implemented. Finally, the effects of these interventions are assessed by measurements of relevant variables.

The rehabilitation process starts at the hospital after the person is medically stable, and all life-threatening conditions are eliminated or reduced. The followings are the major and most common components of rehabilitation programme and often require special attention:

• Rehabilitation nursing:

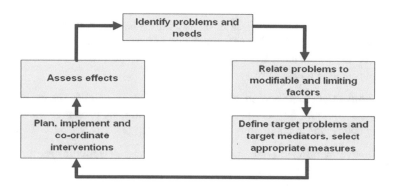

Fig. 2 A modified version of the rehabilitation cycle

- Rehabilitation nursing includes providing care, training and support to the affected individuals and their families. It helps the affected person to develop problem-solving and stress management skills as well as to improve the quality of life of the person through physiological and psychological interventions [10].
- Physical therapy:
- CVM affects the muscle strength and movements of the body or other body parts. Physical therapy aims to improve the muscle strength and mobility, to relieve pain and to retrain for balance and gait through specially designed motor skill exercises, mobility training, constraint-induced (forced-use) therapy, range-of-motion therapy and deep tissue massage [11]. Sometimes heat or cold or ultrasound is used as a part of the therapy.
- Apart from only physical activities, technology-assisted therapies/activities using electricity, robotic devices, wireless technology and virtual reality (VR)/ computer games are also popularly used [11]. The virtual reality– or computer game–based activities can help the affected person to improve strength and coordination between impaired limbs/body and the brain through the interaction with the simulated real-time environment.
- Occupational therapy:
- The occupational therapy helps the affected person to carry out day-to-day activities for independent living [11]. Though the term 'occupation' is related to work or job, but occupational therapy mainly aims to improve the capabilities of the affected person to undertake activities of daily living (ADL), such as taking foods, brushing, toileting, bathing, moving from one location to another, etc. Simultaneously, the knowledge to use walkers, wheelchairs, crutches and other prosthetics are imparted through occupational therapy.
- Speech–language pathology:
- The speech–language pathology addresses to improve disorders related to speech, language and communication. The speech disorder is associated with the difficulty in producing the sounds clearly and fluently [12].
- Audiology recreational therapy:
- Audiology recreational therapy aims to identify, diagnose and manage affected persons with peripheral or central hearing loss, tinnitus and balance disorders.
- Nutritional care:
- It has been found that nutrition or energy intake has a direct relationship with rehabilitation [13]. Nutrition care combined with rehabilitation can provide better recovery of the ADL for the persons affected with CVM.
- Rehabilitation counselling:

- The process of rehabilitation counselling [14] aims to identify the existing disabilities/problems, barriers and potentials of the affected person by the rehabilitation counsellor in collaboration with the affected person.

The above-mentioned components of rehabilitation or activities are important and essential for the person affected with CVM to maximise functionality and to regain the highest possible level of independence. The activities are to be carried out through professionals of respective domains. In the majority of the cases, human professionals can easily carry out the activities, but presently, several technologies are available to assist human professionals. A study of available technologies is presented in the following paragraphs.

3   Available Technological Resources

The shortage of health professionals is a major hindrance in the long term and home-based recovery. The family members, especially the spouse or son or daughter of the affected person, have to take the burden of effective and full-time caregiving. Such caregiving makes the life of the caregiver miserable in terms of physical, physiological, mental, social and financial deterioration. The misery of caregivers can be reduced to some extent by introducing technological assistance and support to the person affected by CVM. Technological resources can be divided into two categories: rehabilitative and assistive technologies. Rehabilitative technologies are directly associated with the improvement process of the lost or affected functions of the disabled persons, whereas assistive technologies can be any product, instrument, equipment or device or aids or tools that are used to increase, maintain or improve functional capabilities of the disabled persons. These two types of technological resources are described in the following paragraphs.

3.1   Rehabilitative Technologies

The rehabilitative technologies are associated directly with the improvement process of the lost or affected functions of the disabled persons. Enhancement of motor skills and mobility is the primary goal of such technologies. Development of such technologies uses the knowledge and help from the field of mechanical, electronic/ electrical and computer engineering and material science. The brief description of such technologies is presented below [15, 16]:

- Video Games – Video games are perhaps the most exciting and popularly used technology for rehabilitation of impaired limbs. The conventional physical activities or the exercises are to be performed repeatedly for better and faster result. But the person affected with CVM may not like to perform such repetitive and monotonous activities, and so improvement is slower. Interactive and exciting video games are used to move objects or move the cursors for triggering some actions. Playing a video game requires physical and cognitive involvements through hand–eye coordination [17]. Often joysticks or joysticks with force feedback (haptics) and Microsoft Kinect Sensor have been used for developing such devices along with cameras, monitors and sensors.
- Robotic Devices or Exoskeleton – Robotic devices or exoskeleton robots are the newest rehabilitative technologies available commercially for the rehabilitation

of impaired limbs. They are directly attached to the affected body parts or limbs, such as legs or hands to facilitate or enable movements of the limbs. L300 Foot Drop System developed by Bioness, Inc. is a commercially available aid that enables a person to move faster with improved balance. WalkAide is a similar aid by Innovative Neurotronics for the rehabilitation of foot drop or drop foot syndrome caused due to CVM. Rehab-Robotics Company Limited has developed the hand of hope therapy device for the rehabilitation of hand and forearm using surface electromyographic (sEMG) signals. Similar rehabilitation device for hand (H200 Wireless Hand Rehabilitation System) has been developed by Bioness, Inc. for helping the person for easy reaching, grasping, opening and closing operations of the hands. The ReoGo by Motorika USA, Inc. is a stationary exoskeleton for arm rehabilitation. This company has also developed a robotic device for high-level gait training, known as ReoAmbulator for improving the balance, ambulation, coordination, stamina, posture, etc. Pneu-WREX is a pneumatic exoskeleton developed by the University of California [22]. Other such similar systems have been presented in Fig. 3a–d below.

- Virtual Reality (VR) – Virtual reality has emerged as a new way of rehabilitation for CVM. The virtual environment is created using various visual interfaces, such as desktop monitors, head-mounted VR system and displays (HMDs) and haptic interfaces along with real-time motion-tracking devices, and the affected person is allowed to interact with virtual objects and images in real time through multiple sensory modalities [23]. Object manipulation and body movement through virtual space in varying degree by the person mimic the situations similar to the real world. Two types of VR system are available for rehabilitation. Immersive VR is created by the equipment/device worn by the user, or the person is situated within the virtual environments. The non-immersive VR is two-dimensional and presented through a computer monitor/screen. The user can control the operations/movements displayed on the screen through joystick or mouse or similar interfacing tools. Figure 4a–c presents a few such VR systems along with their developing companies/universities.
- Functional electrical stimulation (FES) – FES is a rehabilitation technique that uses a small electric pulse to the impaired muscles to restore or improve their function [29]. This electric pulse stimulates the muscle to contract and make its usual movement. FES is generated through a small control box with batteries and

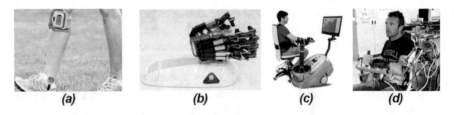

Fig. 3 Robotic and exoskeleton devices for rehabilitation (**a**) L300 Foot Drop System by Bioness, Inc. (**b**) Hand of hope by Rehab-Robotics Company Limited (**c**) ReoGo by Motorika (**d**) Pneu-WREX by the University of California. (Courtesy: [18–21])

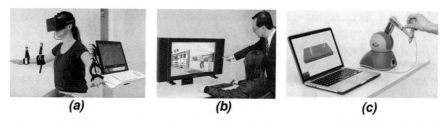

Fig. 4 Virtual reality devices for rehabilitation (**a**) head-mounted VR system (**b**) VR system for the rehabilitation of cognitive deficits (**c**) Geomagic Touch haptic device for upper limb rehabilitation. (Courtesy: [24–28])

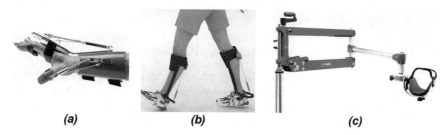

Fig. 5 Passive mechanical devices for rehabilitation (**a**) spring-loaded globes for grasp and release therapy at home (**b**) spring-loaded leg brace (**c**) support system for arms, legs and overall body. (Courtesy: [31])

electrodes. It is very useful for the treatment of foot drop and movement of fingers [30]. The bladder and bowel movements can also be improved using FES.

Apart from the above-mentioned active devices (which are powered and movements between different parts are generated using electricity), several passive mechanical devices have been developed, and some are available commercially for rehabilitation. Majority of these devices use springs, links and grips for generation and controlling the movements of the impaired limbs. A few such selective passive devices have been depicted in Fig. 5a–d.

3.2 Assistive Technologies

Assistive technologies can be any product, instrument, equipment or device or aids or tools that are used to increase, maintain or improve functional capabilities of disabled persons [34]. It can be a simple spoon or fork that has been designed differently (such as they have the provision for attaching it with the body using Velcro or similar rather than holding it or have angled shape than the conventional straight ones) to suit the need of a disabled person or may be a computerised communication device for alternate communication. Assistive technological devices can be so simple that the therapists can even design and develop the low-tech devices themselves for the patients without having much knowledge of engineering.

The uses of assistive technologies are broader. They may be required from the morning to night and even during sleeping also. They may be simpler to the most complex ones in nature. Assistive technologies can be classified based on their domain of use or application:

- Access and Control of the Environment (ACE) – ACE have become easier for persons with disabilities due to advancement and affordability of technologies, especially the wireless and electronics technologies. This has been possible through the use of different alternate input means for accessing and controlling things in the immediate environment, such as switches, special keyboards and mice, wireless control units, etc. to operate one or more electronic appliances for controlling environmental parameters. Presently, such control can even be performed through the apps on mobile phones remotely.

- Simultaneously, high-end technologies have been devised to provide access to easy movements through automatic door opener, faster and stair-climbing attachments or ramps for wheelchairs, etc. Such controlling is either voice-based (known as 'VOICE – ECU') or switch-based ('SWITCH – ECU'). The environmental control units (ECUs) have been shown in Fig. 6a–d below.

- Aids for activities of daily living are required by the disabled persons for independent living and to perform different daily activities, such as brushing, bathing, dressing, eating, transferring to and from beds and many more. The aids for activities of daily living are simpler, based on low-level technologies, and adapt to the environment rather than the person. Figure 7 presents such technologies that support self-care and independent living [33].

- Assistive listening devices are used for overcoming the hearing loss due to CVM. In general, the people with hearing loss require an increase in volume by 15–25 decibels for clarity in hearing. Apart from these, there may be a need for removing background noise for easy and clear hearing for people with hearing

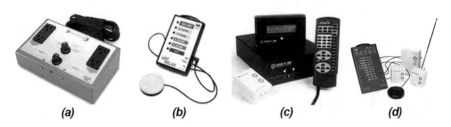

    (a)   *(b)*   *(c)*   *(d)*

Fig. 6 Assistive devices for easy control of the environment (**a**) control unit for operating most electrical appliances, tools, etc. using single switch (**b**) scanning infrared (IR) transmitter to control IR-based devices, such as television (TV), video cassette recorder (VCR), compact disc (CD) player along with pressing switch (**c**) control unit for operating 256 lights and appliances, including televisions, video cassette recorders, compact disc players, electronic bed controls, nurse-call systems and the optional tape recorder (**d**) control unit for operating four IR devices along with ten electrical appliances with a single switch. (Courtesy: [32])

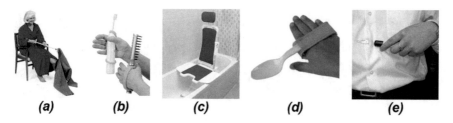

Fig. 7 Aids for different activities of daily living (**a**) dressing hook (**b**) adaptive toothbrush and comb (**c**) bath lift or bathing aids (**d**) universal cuff for utensil holder (**e**) button hook puller. (Courtesy: [36–41])

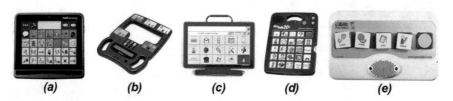

Fig. 8 Augmentative and alternative communication devices (**a**) Tobii Dynavox I-12+ (**b**) MegaBee Assisted Communication and Writing Tablet (**c**) NuEye Tracking System (**d**) Pocket Go-Talk 5-Level Communication Device (**e**) Logan ProxTalker Modular AAC Device. (Courtesy: [42–46])

disability. Amplifiers, frequency modulation (FM) systems, infrared (IR) systems and induction loop systems are some of the assistive listening devices useful for independent living in homes, schools, offices and communities [34, 35].

- Augmentative and alternative communication (AAC) provides alternative methods of communication in the community to supplement or replace writing and speaking for disabled persons with speech and language impairment. AAC devices can range from gestures and sign languages to simple picture boards to a computer programme for synthesising pictures to texts. The unaided AAC does not use any tools or devices, rather uses body or sign languages, gestures, facial expressions, etc. Pencils, picture boards, speech-generation devices, devices producing written outputs and communicator switches are some of the aided AAC devices as presented in Fig. 8a–e below. The selection of AAC device depends upon the type and severity of disability of the affected person.
- Computer access and computer-based instructions are required for impaired persons for easy use of the computer and overcoming the difficulties related to audio-visual output/instruction from the computer. Alternative keyboards, mouse alternatives, touchscreen hardware and software and pointing devices are used to replace the conventional input devices of a computer for easier access as shown in Fig. 9a–d below.
- Mobility is one of the most important issues for the independent living of any person [47]. Mobility is essential for a person affected with CVM to maintain a healthy and hygienic living, performing daily activities of living and socialising

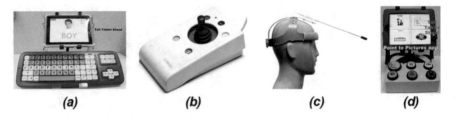

Fig. 9 Assistive arrangements for easy access of the computer (**a**) Bluetooth keyboard with 1-inch keys (**b**) mouse alternative (**c**) head-mounted pointer (**d**) large buttons

Fig. 10 Various assistive technologies (**a**) stair-climber wheelchair for mobility (**b**) adapted chair with desk (**c**) hand-held digital magnifier for low vision (Courtesy: [49–51])

and above all to reduce postural deformities. Mobility impairment can be caused due to the weakness of muscles in the legs or other assisting limbs/body parts, balancing problem and gait disturbances. Mobility aids include simple canes, crutches, walkers, simple and powered wheelchairs, stair-climbing wheelchair and mobility scooters [48]. However, necessary precautions, safety measures and guidance are required for safe movements using mobility aids. Simultaneously, the mobility devices or aids should be lightweight, highly stable, and easy-to-operate.

- Seating and positioning disabilities are overcome by using simple aids, such as seat inserts for wheelchairs, adjustable or adaptive chairs, standers, side-lying frames, head supports, adapted tables/desks, wedges, straps, etc. These aids shown in Fig. 10 can help the disabled person to remain in a good position without any pain or tiredness.
- Visual aids are required for the person affected with the loss of vision or impaired vision due to CVM. Vision loss is a very common after-effect of CVM, and access to information is essential by any person. So, assistance may be required to access or read written information and produce information in writing. Visual devices or aids may include the followings but not limited to contrast enhancer; image magnifier; tactile and auditory materials; recorded books (on tape); books wither larger prints; aids for low vision; screen magnifier; screen magnification software; electronic magnifier; screen reader; Braille keyboard; Braille translator

Fig. 11 A low-cost eye-tracking system for the rehabilitation of the completely locked-in patient (**a**) hardware set-up along with optimum position of the eyes (**b**) graphical user interface (GUI) representing various basic daily needs of human and the special button to 'enable' or 'disable' the buttons

    software; Braille printer/embosser; optical character readers; or text-to-speech converting machine.
- The preceding discussion reveals the components of rehabilitation for physical, cognitive and emotional improvements towards self-independence after CVM. Human and technological resources are the two components for effective and successful rehabilitation. The majority of the rehabilitation aids/systems focus on the improvements of neuromotor impairments. Costlier and advanced technologies have been developed for such purpose. But improvements of neuro-motor impairments cannot alone make an affected person independent unless his quality of life is improved simultaneously.
- Performing daily activities independently by the CVM-affected people is very important part of rehabilitation. They need to communicate and express their desires to perform activities of daily living (ADL) independently or with assistance during any time of the day. The situation becomes critical if the affected person with physical impairment loses memory and speech.

Only a few kinds of research have been reported for the development of such rehabilitation system using icons, infrared (IR) lights and wearable eye trackers. The first idea of such technology was proposed in 2006 by Abu-Faraj et al. [52]. For the development of the said technology, a low-cost webcam and a laptop have been used to provide an alternative means of communication for CVM patients, as shown in Fig. 11a.

Another similar device has been developed for communication for people with severe muscular disability and speech impairment [53]. A low-cost webcam that has been modified to work as an IR camera is mounted on the head of the user using either a helmet or eyeglasses as depicted in Fig. 12a, b, respectively.

4 Conclusions

Cerebrovascular malfunction affects the blood vessels inside the brain, leading to various neurological, cognitive and motor disorders. The most common post-CVM effects are limited or lack of mobility of limbs or body and difficulties in

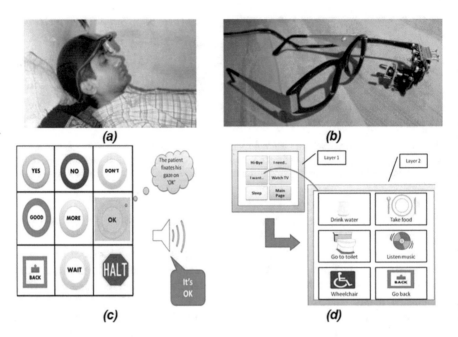

Fig. 12 Various components (hardware and the display screens) of an eye-gaze-based communication device for people with disability (**a**) helmet-mounted camera (**b**) camera mounted on eyeglasses (**c**) single-layered display screen (**d**) two-layered approach for menu-driven selections

communication due to the loss of memory, language problem, affected thought process, etc. Initially, rehabilitation plays a vital role to regain self-independence of the affected person, and later, proper assistance is to be provided for carrying out daily activities. Computer hardware and software; electronic devices and systems; mechanical devices and aids; and, above all, material science are the essential ingredients of such technologies. Video games, virtual reality, robotic systems and exoskeleton devices are useful for the rehabilitation of the affected persons. Assistive technologies are the simple or complex devices or mechanisms or means used to access and control the environment and to carry out the activities of daily living, listening, communicating, movement, seating and positioning, visualising, etc.

Due to the lack of trained health professionals, increase in aged people worldwide and with the increase in technologies and innovation, the domain is expanding day by day. The present work focused on the developed technologies for rehabilitation and assistance. Furthermore, the research gaps are being identified, and attempts will be made to fulfil those gaps through the development of affordable and innovative technologies.

Acknowledgments The authors are grateful to Dr. D. N. Ray for his continuous suggestions and advices, without which this work would not have been completed.

References

1. W. Johnson, O. Onuma, M. Owolabi, S. Sachdev, Stroke: a global response is needed. Bull. World Health Organ. **94**, 634–634A (2016). https://doi.org/10.2471/BLT.16.181636
2. World Health Organization, *Rehabilitation* [online]. Available from: www.who.int/disabilities/world_report/2011/chapter4.pdf (2011)
3. H. Haghgoo, E. Pazuki, A. Hosseini, M. Rassafiani, Depression, activities of daily living and quality of life in patients with stroke. J. Neurol. Sci. **328**(1–2), 87–91 (2013)
4. L. Brewer, F. Horgan, A. Hickey, D. Williams, Stroke rehabilitation: recent advances and future therapies. Int. J. Med. **106**(1), 11–25 (2013). https://doi.org/10.1093/qjmed/hcs174
5. B.H. Dobkin, Rehabilitation after stroke. N. Engl. J. Med. **352**(16), 1677–1684 (2005). https://doi.org/10.1056/NEJMcp043511
6. N. Gupta, C. Castillo-Laborde, M.D. Landry, Health-related rehabilitation services: assessing the global supply of and need for human resources. BMC Health Serv. Res. **11**, 276 (2011)
7. World Health Organization, *The World Health Report 2006 – Working Together for Health* [Online]. Available from: https://www.who.int/whr/2006/overview_fig3_en.pdf (2006)
8. World Health Organization, *The Need to Scale Up Rehabilitation: Rehabilitation 2030 – A Call for Action* [Online]. Available from: https://www.who.int/disabilities/care/NeedToScaleUpRehab.pdf (2017)
9. E.D. Ona, P. Baeza, H.A. JARDON, C. Balaguer, Review of Automated Systems for Upper Limbs Functional Assessment in Neurorehabilitation. IEEE Access **1**(1), 99 (2019). https://doi.org/10.1109/ACCESS.2019.2901814
10. A. Koç, Rehabilitation nursing: applications for rehabilitation nursing. Int. J. Caring Sci. **5**(2), 80–86 (2012)
11. National Institute of Health, *Rehabilitation Medicine*. https://www.cc.nih.gov/rmd/ (2018)
12. M. Stephens, The Effectiveness of Speech and Language Therapy for Poststroke Aphasia. Am. J. Nurs. **117**(11), 19 (2017). https://doi.org/10.1097/01.NAJ.0000526741.00314.d9
13. M. Nii, K. Maeda, H. Wakabayashi, S. Nishioka, A. Tanaka, Nutritional improvement and energy intake are associated with functional recovery in patients after cerebrovascular disorders. J. Stroke Cerebrovasc. Dis. **25**(1), 57–62 (2016). https://doi.org/10.1016/j.jstrokecerebrovasdis.2015.08.033
14. F. Chan, J. Chronister, D. Catalana, A. Chase, L. Eun-Jeong, Foundations of rehabilitation counseling. Direct. Rehabil. Counsel. **15**, 1–11 (2004)
15. C. Winstein, P. Requejo, Innovative technologies for rehabilitation and health promotion: what is the evidence? Phys. Ther. **95**(3), 294–298 (2015). https://doi.org/10.2522/ptj.2015.95.2.294.
16. K.S.G. Chua, C.W.K. Kuah, Innovating with rehabilitation technology in the real world promises, potentials, and perspectives. Am. J. Phys. Med. Rehabil. **96**(10), S150–S156 (2017). https://doi.org/10.1097/PHM.0000000000000799
17. K. Lohse, N. Shirzad, A. Verster, N. Hodges, H.F. Van der Loos, Video games and rehabilitation: using design principles to enhance engagement in physical therapy. J. Neurol. Phys. Ther. **37**(4), 166–175 (2013)
18. J. Hausdorf, H. Ring, The effect of the NESS L300 neuroprosthesis on gait stability and symmetry. J. Neurol. Phys. Ther. **30**(4), 198 (2006). https://doi.org/10.1097/01.NPT.0000281266.34830.4b
19. F. Aggogeri, T. Mikolajczyk, J. O'Kane, Robotics for rehabilitation of hand movement in stroke survivors. Adv. Mech. Eng. **11**(4), 1–14 (2019). https://doi.org/10.1177/1687814019841921
20. T. Takebayashi, K. Takahashi, S. Amano, Y. Uchiyama, M. Gosho, K. Domen, K. Hachisuka, Assessment of the efficacy of ReoGo™-J robotic training against other rehabilitation therapies for upper-limb hemiplegia after stroke: protocol for a randomized controlled trial. Front. Neurol. **9**, 730 (2018)
21. D.J. Reinkensmeyer, E.T. Wolbrecht, V. Chan, C. Chou, S.C. Cramer, J.E. Bobrow, Comparison of three-dimensional, assist-as-needed robotic arm/hand movement training provided with

Pneu-WREX to conventional tabletop therapy after chronic stroke. Am. J. Phys. Med. Rehabil. **91**(11), S232–S241 (2012). https://doi.org/10.1097/PHM.0b013e31826bce79

22. R. Secoli, M. Milot, G. Rosati, D. Reinkensmeyer, Effect of visual distraction and auditory feedback on patient effort during robot-assisted movement training after stroke. J. Neuroeng. Rehabil. **8**(1), 1–10 (2011)

23. H. Sveistrup, Motor rehabilitation using virtual reality. J. Neuroeng. Rehabil. **1**(1), 10–18 (2004)

24. W.J. Shaun, G. Mark, C.V.W. Hugo, N.K.B. Maged, Head-mounted virtual reality and mental health: critical review of current research. JMIR Serious Games **6**(3), e14 (2018). https://doi.org/10.2196/games.9226

25. Hong Kong Polytechnic University, *Virtual-Reality-Based Rehabilitation System Helps People with Cognitive Deficits* [Online]. Available from: http://www.polyu.edu.hk/cpa/polyumilestones/14Jun/pdf/201412_PolyU_Milestones.pdf (2014)

26. E. Klinger, J. Sánchez, P. Sharkey, J. Merricket, Virtual reality-based rehabilitation applications for motor, cognitive and sensorial disorders. Int. J. Disabil. Human Dev. **13**(3), 309–309 (2014). https://doi.org/10.1515/ijdhd-2014-0320

27. A.L. Faria, A. Andrade, L. Soares, S.B. Badia, Benefits of virtual reality based cognitive rehabilitation through simulated activities of daily living: a randomized controlled trial with stroke patients. J. NeuroEng. Rehabil. **13**, Article number: 96 (2016)

28. L. Piggot, S. Wagner, M. Ziat, Haptic neurorehabilitation and virtual reality for upper limb paralysis: a review. Critic. Rev. Biomed. Eng. **44**(1). Available from: 10.1615/CritRevBiomedEng.2016016046 (2016)

29. Y. Hara, Rehabilitation with Functional Electrical Stimulation in Stroke Patients. Int J Phys Med Rehabil. **1**, 147 (2013). https://doi.org/10.4172/2329-9096.1000147

30. D.N. Rushton, Functional electrical stimulation and rehabilitation— a hypothesis. Med. Eng. Phys. **25**(1), 75–78 (2003)

31. B. Doucet, J. Mettler, Pilot study combining electrical stimulation and a dynamic hand orthosis for functional recovery in chronic stroke. Am. J. Occup. Ther. **72**(2) (2018)

32. AbleNet, Inc., Environmental Control Units (ECU) Catalog (2008)

33. J. Gierach, K. Stindt, Assistive technology for activities of daily living, in *Assessing Students' Needs for Assistive Technology*, (Wisconsin Assistive Technology Initiative, 2009)

34. H. Blum, The assistive-tech future is hear. ASHA Leader (2017). https://doi.org/10.1044/leader.FTR2.22022017.52

35. A.M. Rekkedal, Assistive hearing technologies among students with hearing impairment: factors that promote satisfaction. J. Deaf Stud. Deaf Educ. **17**(4), 499–517 (2012). https://doi.org/10.1093/deafed/ens023

36. H. Smith, *The 7 Best Daily Living Aids* [Online]. Available from: https://www.rehabmart.com/best-products/best-daily-living-aids (2018)

37. J. Ripat, Function and impact of electronic aids to daily living for experienced users. Technol. Disabil. **18**(2), 79–87 (2006). https://doi.org/10.3233/TAD-2006-18204

38. P. Rigby, S. Ryan, S. Joos, B. Cooper, J. Jutai, I. Steggles, Impact of electronic aids to daily living on the lives of persons with cervical spinal cord injuries. Assist. Technol. Off. J. RESNA **17**(2), 89–97 (2005). https://doi.org/10.1080/10400435.2005.10132099

39. M.C. Verdonck, G. Chard, M. Nolan, Electronic aids to daily living: be able to do what you want. Disabil. Rehabil. Assist. Technol. **6**(3), 268–281 (2011). https://doi.org/10.3109/17483107.2010.525291

40. I.-L. Boman, K. Tham, A. Granqvist, A. Bartfai, H. Hemmingsson, Using electronic aids to daily living after acquired brain injury: a study of the learning process and the usability. Disabil. Rehabil. Assist. Technol **2**(1), 23–33 (2007). https://doi.org/10.1080/17483100600856213

41. M.L. Lange, The future of electronic aids to daily living. Am. J. Occup. Ther. Off. Publ. Am. Occup.l Ther. Assoc. **56**(1), 107–109 (2002)

42. Tobii Technology, Tobii Dynavox Devices (2015)

43. T. Caudill, *MegaBee Electronic Hand-held Writing Tablet* (VA Assistive Technology, Pittsburg, 2017)

44. Prentke Romich Company, Nu.Eye Tracking System: User's Guide (2012)
45. K.M. Tonsing, S. Dada, Teachers' perceptions of implementation of aided AAC to support expressive communication in South African special schools: a pilot investigation. Augment. Altern. Commun. **32**(4), 282–304 (2016). https://doi.org/10.1080/07434618.2016.1246609
46. Communication Matters, The Journal of Communication Matters, 29 (2), International Society for Augmentative and Alternative Communication, UK (2015)
47. C.C. Caro, J.D. Costa, D.M.C. da Cruz, The use of mobility assistive devices and the functional independence in stroke patients. Cadernos Brasileiros de Terapia Ocupacional **26**(3), 558–568 (2018). https://doi.org/10.4322/2526-8910.ctoAO1117
48. R. Cowan, E. Fregly, B.J. Boninger, M.L. Chan, L. Rodgers, M.M. Reinkensmeyer, D. J., Recent trends in assistive technology for mobility. J. NeuroEng. Rehabil. **9**(20), 1–8 (2012)
49. J. Peskett, Stanley Handling to tackle stairlift sector head-on. Access and mobility professional (2017)
50. J. Karen, R. Penny, O. Marcia, Assistive technology for positioning, seating, and mobility, in *Assessing Students' Needs for Assistive Technology*, (WATI, 2009)
51. G. Virgili, R. Acosta, L.L. Grover, S.A. Bentley, G. Giacomelli, Reading aids for adults with low vision. Cochrane Database Syst. Rev. **10**(1), CD003303 (2013). https://doi.org/10.1002/14651858.CD003303.pub3
52. Z.O. Abu-Faraj, M.J. Mashaalany, H.C.B. Sleiman, J.D. Heneine, W.M.A. Katergi, Design and development of a low-cost eye tracking system for the rehabilitation of the completely locked-in patient, in *2006 International Conference of the IEEE Engineering in Medicine and Biology Society, 30 Aug.-3 Sept. 2006*, (IEEE, New York, 2006), pp. 4905–4908
53. A. Srivastava, A. Asati, *Eye-Gaze based Communication Device for People with Disability* (Maulana Azad National Institute of Technology, Bhopal, 2012)

Machine Learning Application: Sarcasm Detection Model

Swati Sharma, Vijay Kumar Sharma, Vimal Kumar, and Umang Arora

1 Introduction

The combination of negative feelings with positive words is termed as sarcasm. It is
something that defines the distinction between one's mind and tongue. Sarcasm can
be taken in two forms, as criticism or fun. It is an analyst's job to examine the literal
interpretation of the sentence as well as the context in which the sentence has been
stated to identify the underlying sarcasm. With the use of emoticons, capital letters,
exclamation marks, or slang [1], sarcasm is more ubiquitous and apparent. To detect
the presence of sarcasm, we have developed and presented, in this chapter, a model
based on artificial intelligence. Although irony, satire, and sarcasm appear almost
same, there is an appropriate difference among them. The irony is cited as a gap
between expectations and reality, satire is embellishment, and sarcasm is an art of
mockery [2, 3].

Example: "This topic making me lunatic."

In state of sarcasm:
Consider a candidate. When a topic is asked from him and he doesn't know its
answer and later he replies, "This topic making me lunatic," it conveys not to ask
such kind of questions.

In state of irony:
Consider a candidate who is extremely confident that he knows all the answers.
When a topic is asked from him and he doesn't know its answer and later he

S. Sharma (✉)
Department of Information Technology, MIET, Meerut, India
e-mail: swati.sharma.it@miet.ac.in

V. K. Sharma · V. Kumar · U. Arora
Department of Computer Science and Engineering, MIET, Meerut, India
e-mail: vijay.sharma@miet.ac.in; vimal.kumar@miet.ac.in; umang.arora@miet.ac.in

replies, "This topic making me lunatic," it conveys the irony as the expectations and reality didn't match.

In state of satire:
Consider a candidate who never takes any stress while answering any question. When a topic is asked from him and he doesn't know its answer and later he replies, "This topic making me lunatic," it conveys satire as he was taking stress to answer.

2   Research Methodology

The iterative flowchart of the proposed methodology is depicted in Fig. 1. The initial step is parsing the dataset into a Pandas DataFrame. The presence of null values is checked and removed. The next step is data cleaning. The process of correcting or

Fig. 1 Flowchart for sarcasm detection model

removing corrupt or inaccurate records and fields from the dataset is called data cleaning. We reduce the number of features from the existing ones using the feature extraction technique [4]. After the useful features are extracted, we use stemming algorithm to produce morphological variants of the root words. Term frequency–inverse document frequency (TF-IDF) is then performed to transform the text into meaningful numerical representation to fit machine learning algorithm [5].

After setting the model, the model is ready for training and testing. Last, we set up the GUI application using Tkinter, it provides a very powerful and an object-oriented interface for the GUI tool kit, and finally, the results are displayed [6, 7].

2.1 Parsing Dataset into DataFrames

We have elucidated a dataset containing sentences characterized into sarcastic and nonsarcastic and loaded the dataset into Pandas DataFrame. A DataFrame is a data structure with two dimensions used to store labeled data with columns of potentially different types [8].

```
>>> dataset = pd.read_json('SARCAM_DATASET.json', lines = True)
```

The dataset comprises the following headers:

1. article_link: to hold link of news articles
2. headline: to hold headline of news
3. is_sarcastic: to hold 0 for nonsarcastic and 1 for sarcastic.

Table 1 represents that the sentence is sarcastic or not according to the used dataset.

2.2 Checking Null Values

The null values are tested in the dataset. If the dataset doesn't contain any null value, it can be further processed [9]. Figure 2 shows the output data.

```
>>> print (data.isnull (). any (axis = 0))
```

2.3 Data Cleaning

In data cleaning, uninformative tweets, redundant tweets, punctuations, emoticons, and uniform resource locators (URLs) are removed. Then, data parsing is being done, in which non-American Standard Code for Information Interchange (ASCII)

Table 1 Sample of dictionary representing sarcasm or nonsarcasm

S. No	article_link	headline	is_sarcastic
1	https://www.huffingtonpost.com/entry/ bloomberg-philanthropies-what-works-cities-expands_us_566746f3e4b080eddf55ee73	Bloomberg's program to build better cities just got bigger	0
2	https://local.theonion.com/ courtroom-sketch-artist-has-clear-manga-influences-1820298494"ml	Courtroom sketch artist has clear manga influences.	1
3	https://www.huffingtonpost.com/entry/ qatar-dutch-woman-raped_us_575eb891e4b00f9 7fba8cead	Qatar deporting Dutch woman who reported she was drugged and raped	0
4	https://www.huffingtonpost.com/entry/ obama-veterans-day_us_564372e9e4b08cda3486 f09b	Obama visits Arlington National Cemetery to honor veterans	0
5	https://www.huffingtonpost.com/entry/ remembrance-is-the-beginn_b_5382344.html	Remembrance is the beginning of the task.	0
6	https://www.theonion.com/ stock-analysts-confused-frightened-by-boar-market-1819567580	Stock analysts confused, frightened by boar market	1

Fig. 2 Output

```
---------------------------------------------
Article_link               False

Headline                   False

Is_sarcastic               False

Dtype: bool
---------------------------------------------
```

hashtags are removed [10, 11]. After removing the null values, the special symbols are eliminated by using regular expressions. Figure 3 shows the process of data cleaning.

2.4 Feature and Label Extraction

In the feature level analysis, the opinions are identified on the dependency of extracted features from document level and sentence level analyses. Here, we have considered features as headline and label as is_sarcastic [12]. The process of feature extraction and label extraction is shown in Fig. 4.

```
dataset['headline']  = dataset['headline'].
        apply(lambda s : re.sub('[^a-zA-Z]', ' ' , s))
```

Fig. 3 Data cleaning

Fig. 4 Feature and label extraction

```
features = dataset['headline']
labels = dataset['is_sarcastic']
```

```
ps = PorterStemmer()
features = features.apply (lambda x: x.split())
features = features.apply (lambda x: ' '.join(
                    [ps.stem(word) for word in x]))
```

Fig. 5 Stemming

2.5 Stemming

Stemming is a procedure of reducing derived words. The stem may not necessarily be the same as the morphological root word. Here, we have used Porter stemmer for stemming process [13, 14]. Figure 5 shows the code snippet for carrying out stemming.

2.6 TF-IDF Vectorization

Term frequency–inverse document frequency is a vectorization algorithm used for transforming the text into meaningful depiction of numbers [15]. Figure 6 shows the vectorization of the extracted features.

2.7 Training and Testing of Data

Training and testing involves converting the list into NumPy arrays; after converting, the model is trained and then the testing is done for checking the accuracy of the trained model [16, 17]. Training and testing is carried out using code depicted in Fig. 7.

Figure 8 shows the sample of training and testing with train_loss, train_accuracy, test_loss, and test_accuracy.

```
from sklearn.feature_extraction.text import TfidVectorizer
vec = TfidVectorizer (max_features = 5000)
features = list(features)
features = vec.fit_transform(features).toarray()
```

Fig. 6 TF-IDF Vectorization

```
training_es = tokenizer.texts_to_sequences(training_sentences)
training_padded = pad_sequences(training_sequences, maxlen=max_length,
                            padding=padding_type, truncating=trunc_type)
testing_sequences = tokenizer.texts_to_sequences(testing_sentences)
testing_padded = pad_sequences(testing_sequences, maxlen=max_length,
                            padding=padding_type, truncating=trunc_type)
```

Fig. 7 Training and testing data

```
Epoch 38/50                                            ̄                    ̄
626/626 - 2s - loss: 0.0153 - accuracy: 0.9961 - val_loss: 1.3122 - val_accuracy: 0.8068
Epoch 39/50
626/626 - 2s - loss: 0.0147 - accuracy: 0.9959 - val_loss: 1.3547 - val_accuracy: 0.8077
Epoch 40/50
626/626 - 2s - loss: 0.0134 - accuracy: 0.9960 - val_loss: 1.3964 - val_accuracy: 0.8078
Epoch 41/50
626/626 - 2s - loss: 0.0119 - accuracy: 0.9966 - val_loss: 1.4160 - val_accuracy: 0.8081
Epoch 42/50
626/626 - 2s - loss: 0.0117 - accuracy: 0.9967 - val_loss: 1.4455 - val_accuracy: 0.8040
Epoch 43/50
626/626 - 2s - loss: 0.0100 - accuracy: 0.9976 - val_loss: 1.4856 - val_accuracy: 0.8035
Epoch 44/50
626/626 - 2s - loss: 0.0093 - accuracy: 0.9976 - val_loss: 1.5287 - val_accuracy: 0.8037
Epoch 45/50
626/626 - 2s - loss: 0.0102 - accuracy: 0.9972 - val_loss: 1.6048 - val_accuracy: 0.8058
Epoch 46/50
626/626 - 2s - loss: 0.0093 - accuracy: 0.9976 - val_loss: 1.6420 - val_accuracy: 0.8046
Epoch 47/50
626/626 - 2s - loss: 0.0087 - accuracy: 0.9977 - val_loss: 1.6384 - val_accuracy: 0.8049
Epoch 48/50
626/626 - 2s - loss: 0.0091 - accuracy: 0.9978 - val_loss: 1.6701 - val_accuracy: 0.8032
Epoch 49/50
626/626 - 2s - loss: 0.0083 - accuracy: 0.9977 - val_loss: 1.7122 - val_accuracy: 0.8053
Epoch 50/50
626/626 - 2s - loss: 0.0068 - accuracy: 0.9981 - val_loss: 1.7495 - val_accuracy: 0.8040
```

Fig. 8 Glimpse of training

2.8 Training and Testing of the Model

After training and testing of data, the model is trained using different machine learning algorithms such as linear support vector classifier, Gaussian Naive Bayes, logistic regression, and random forest classifier [18]. Figure 9 shows the training and testing of the model based on linear support vector classifier (LSVC) algorithm. LSVC is a supervised learning algorithm, which is used for classification as well as

```
#  Algorithm 1
#  Using Linear Support Vector Classifier LSVC
lsvc = LinearSVC()
# Model Training
svc.fit(features_train, labels_train)
# Score of Train and Test data
print(svc.score(features_train, labels_train))
print(svc.score(features_test, labels_test))
```

Fig. 9 Model based on linear support vector classifier

```
#  Algorithm 2
#  Using Gaussuan Naive Bayes NB
nb = GaussianNB()
nb.fit(features_train, labels_train)
print(nb.score(features_train, labels_train))
print(nb.score(features_test, labels_test))
```

Fig. 10 Model based on Gaussian Naive Bayes algorithm

```
#  Algorithm 3
#  Logistic Regression
lor = LogisticRegression()
lor.fit(features_train, labels_train)
print(lor.score(features_train, labels_train))
print(lor.score(features_test, labels_test))
```

Fig. 11 Model based on logistic regression algorithm

regression problems that create a decision boundary that can segregate n-dimensional space into classes [19, 20].

Figure 10 shows the training and testing of the model based on Gaussian Naive Bayes algorithm. Naive Bayes are supervised machine learning algorithms based on the Bayes theorem. Gaussian Naive Bayes follows Gaussian normal distribution and is a Naive Bayes variant that supports continuous data [21].

Figure 11 shows the training and testing of the model based on logistic regression algorithm. Logistic regression is a supervised learning classification algorithm used to predict the probability of a dichotomous target variable [22].

Figure 12 shows the training and testing of the model based on random forest classifier. Using random decision forest classifiers is a method of classification that works by constructing multiple decision trees at the training time and outputting the class that is the mode or mean of the individual trees [23].

```
# Algorithm 4
# Random Forest Classifier
rafc = RandomForestClassifier(n_estimators = 10, random_state = 0)
rafc.fit(features_train, labels_train)
print(rafc.score(features_train, labels_train))
print(rafc.score(features_test, labels_test))
```

Fig. 12 Model based on random forest classifier

Table 2 Sample of dictionary representing sarcasm or nonsarcasm

S. No.	Algorithm	Training%	Testing%
1	Linear support vector Classifier	84	91
2	Gaussian Naive Bayes	74	79
3	Logistic regression	83	88
4	Random forest classifier	80	99

Table 2 shows the training and testing percentages of different algorithms.

The graphical representation of the training and testing accuracy of each of the four models is shown in Figs. 13 and 14, respectively.

2.9 Setting up GUI Application Using Tkinter

Tkinter is an object-oriented layer over Tcl/Tk. Tk is not section of Python; it is prolonged at the active state. It is an interface for Tk GUI toolkit embedded with Python. Python when entangled with Tkinter bestows with the fastest and an easier method for the creation of GUI application. It provides a very powerful, an object-oriented interface for the Tk GUI tool kit. Figure 15a, b shows the screenshots of the developed model.

3 Results and Discussion

Table 3 shows the training accuracy and testing accuracy at different epochs. In a total of 50 epochs, the developed model has received its highest accuracy as 99.39%.

The graph in Fig. 16 shows the training accuracy at different epochs. Figure 17 shows the testing accuracy at different epochs, and Fig. 18 shows the comparison between training and testing accuracy.

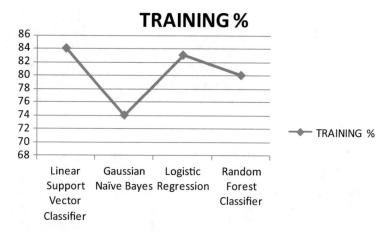

Fig. 13 Accuracy on training

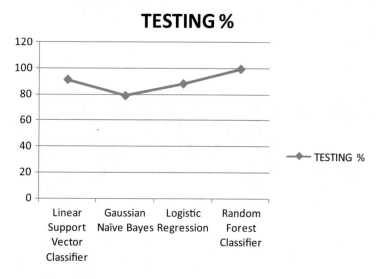

Fig. 14 Accuracy on testing

4 Conclusion

To detect the presence of sarcasm in a given sentence, we have developed machine learning model based on linear support vector classifier (LSVC) algorithm. The training and testing of data was performed, and the highest train and test accuracy was achieved by LSVC model that produced the state-of-the-art accuracy of 99.81% on training and 80.4% accuracy on test set in 50 epochs. In a future work, an analysis on irony, satire, and other literary devices in textual and speech can be carried out.

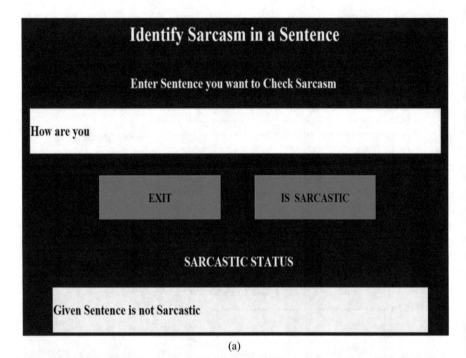

(a)

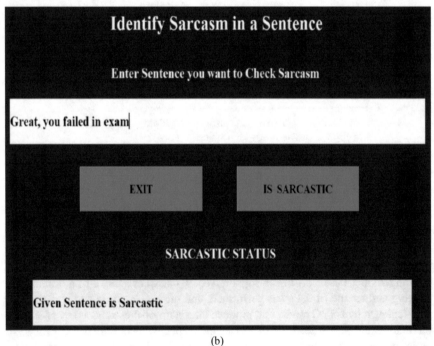

(b)

Fig. 15 (**a, b**) Screenshot of the developed model

Table 3 Training and testing accuracy at different epochs

EPOCH	ACCURACY	VAL_ACCURACY
0	0.5818	0.7517
1	0.8250	0.8384
2	0.8733	0.8424
3	0.8953	0.8519
4	0.9110	0.8552
5	0.9215	0.8430
6	0.9312	0.8411
7	0.9392	0.8549
8	0.9449	0.8544
9	0.9520	0.8531
10	0.9557	0.8517
11	0.9600	0.8468
12	0.9639	0.8487
13	0.9667	0.8291
14	0.9715	0.8457
15	0.9746	0.8383
16	0.9770	0.8378
17	0.9773	0.8365
18	0.9789	0.8348
19	0.9822	0.8323
20	0.9836	0.8294
21	0.9852	0.8282
22	0.9854	0.8231
23	0.9873	0.8231
24	0.9873	0.8222
25	0.9890	0.8228
26	0.9900	0.8206
27	0.9914	0.8191
28	0.9915	0.8180
29	0.9924	0.8171
30	0.9907	0.8153
31	0.9937	0.8143
32	0.9936	0.8132
33	0.9941	0.8117
34	0.9951	0.8108
35	0.9943	0.8093
36	0.9958	0.8049
37	0.9961	0.8068
38	0.9959	0.8077
39	0.9960	0.8078
40	0.9966	0.8081
41	0.9967	0.8040
42	0.9976	0.8035
43	0.9976	0.8037

(continued)

Table 3 (continued)

EPOCH	ACCURACY	VAL_ACCURACY
44	0.9972	0.8058
45	0.9976	0.8046
46	0.9977	0.8049
47	0.9978	0.8032
48	0.9977	0.8053
49	0.9981	0.8040

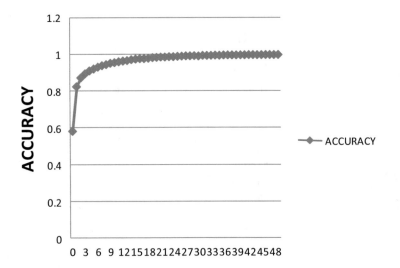

Fig. 16 Training accuracy

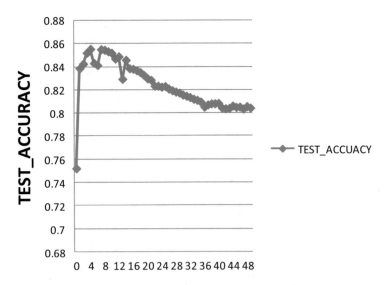

Fig. 17 Testing accuracy

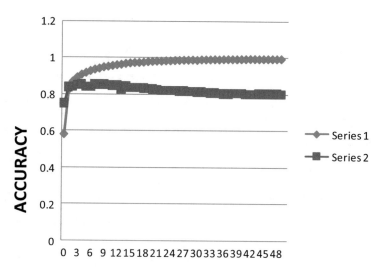

Fig. 18 Comparison between training and testing accuracy

Acknowledgments We would like to thank God for helping us in every condition.

References

1. P.N. Howard, M.M. Hussain, The upheavals in Egypt and Tunisia: The role of digital media. J. Democracy **22**(3), 35–48 (2011)
2. N.A. Diakopoulos, D.A. Shamma, Characterizing debate performance via aggregated twitter sentiment, in *Proceedings of the SIGCHI Conference on Human Factors in Computing Systems*, (2010), pp. 1195–1198
3. O. Kucuktunc, B.B. Cambazoglu, I. Weber, H. Ferhatosmanoglu, Characterizing debate performance via aggregated twitter sentiment, in *Proceedings of the SIGCHI Conference on Human Factors in Computing Systems*, (2010)
4. S.W. Cho, M.S. Cha, S.Y. Kim, J.C. Song, K.A. Sohn, Investigating temporal and spatial trends of brand images using twitter opinion mining, in *International Conference on Information Science & Applications (ICISA)*, (IEEE, 2014), pp. 1–4
5. A. Kumar, A. Jaiswal, Systematic literature review of sentiment analysis on Twitter using soft computing techniques. Concurr. Comput. Pract. Exp. **32**(1), e5107 (2020)
6. H. Fujita, Neural-fuzzy with representative sets for prediction of student performance. Appl. Intell. **49**(1), 172–187 (2019)
7. Y.H. Robinson, E.G. Julie, K. Saravanan, R. Kumar, FD-AOMDV: fault-tolerant disjoint ad-hoc on-demand multipath distance vector routing algorithm in mobile ad-hoc networks. J. Ambient Intell. Human. Comput. **10**(11), 4455–4472 (2019)
8. K. Saravanan, A. Selva, K. Raghvendra, How to prevent maritime border collision for fisheries?-A design of Real-Time Automatic Identification System. Earth Sci. Inform. **12**(2), 241–252 (2019)
9. N.T. Tam, D.T. Hai, Improving lifetime and network connections of 3D wireless sensor networks based on fuzzy clustering and particle swarm optimization. Wireless Netw. **24**(5), 1477–1490 (2018)

10. N.D. Thanh, M. Ali, A novel clustering algorithm in a neutrosophic recommender system for medical diagnosis. Cogn. Comput. **9**(4), 526–544 (2017)
11. A.K. Jena, A. Sinha, R. Agarwal, C-net: contextual network for sarcasm detection, in *Proceedings of the Second Workshop on Figurative Language Processing*, (2020), pp. 61–66
12. A. Joshi, P. Bhattacharyya, M.J. Carman, Automatic sarcasm detection: a survey. ACM Comput. Surv. (CSUR) **50**(5), 1–22 (2017)
13. M. Bouazizi, T.O. Ohtsuki, A pattern-based approach for sarcasm detection on twitter. IEEE Access **4**, 5477–5488 (2016)
14. A. Joshi, V. Tripathi, K. Patel, P. Bhattacharyya, M. Carman, Are word embedding-based features useful for sarcasm detection? arXiv preprint, arXiv:1610.00883 (2016)
15. T. Ptáček, I. Habernal, J. Hong, Sarcasm detection on Czech and English twitter, in *Proceedings of COLING 2014, the 25th International Conference on Computational Linguistics: Technical Papers*, (2014), pp. 213–223
16. A. Joshi, V. Sharma, P. Bhattacharyya, Harnessing context incongruity for sarcasm detection, in *Proceedings of the 53rd Annual Meeting of the Association for Computational Linguistics and the 7th International Joint Conference on Natural Language Processing*, Volume 2: Short Papers, (2015), pp. 757–762
17. A. Rajadesingan, R. Zafarani, H. Liu, Sarcasm detection on twitter: a behavioral modeling approach, in *Proceedings of the Eighth ACM International Conference on Web Search and Data Mining*, (2015), pp. 97–106
18. S. Amir, B.C. Wallace, H. Lyu, P.C.M.J. Silva, Modelling context with user embeddings for sarcasm detection in social media. arXiv preprint, arXiv:1607.00976 (2016)
19. A. Ghosh, T. Veale, Magnets for sarcasm: making sarcasm detection timely, contextual and very personal, in *Proceedings of the 2017 Conference on Empirical Methods in Natural Language Processing*, (2017), pp. 482–491
20. D. Hazarika, S. Poria, S. Gorantla, E. Cambria, R. Zimmermann, R. Mihalcea, Cascade: Contextual sarcasm detection in online discussion forums. arXiv preprint, arXiv:1805.06413 (2018)
21. Lunando, Edwin, and Ayu Purwarianti.: Indonesian social media sentiment analysis with sarcasm detection. In: International Conference on Advanced Computer Science and Information Systems (ICACSIS), IEEE (2013) 195-198.
22. Z. Wang, Z. Wu, R. Wang, Y. Ren, Twitter sarcasm detection exploiting a context-based model, in *International Conference on Web Information Systems Engineering*, (Springer, Cham, 2015), pp. 77–91
23. A. Mishra, D. Kanojia, S. Nagar, K. Dey, P. Bhattacharyya, Harnessing cognitive features for sarcasm detection. arXiv preprint, arXiv:1701.05574 (2017)

Modern Technology on Building Marketing 4.0: Impact on Customer Engagement

Tripti Sharma, Arvind Kumar Bhatt, and Amare Abawa

1 Introduction

1.1 About Digital and Social Media Emergence

According to Liao, [11] the development of new digital technologies and techniques tends to be pronounced as the Fourth Industrial Revolution (Industry 4.0). Industry 4.0 and smart commerce are related expressions, often used identically. The concept of Industry 4.0 denotes the digital transformation of entire value chains, business functional processes, business models, products, and service offerings. The entire universe has been progressively changing over the past decades due to the government support, increasing awareness among millennial generation, consumer orientation, and fast fruition of information technologies. Market-oriented approach and marketing practices are also shifting at the same pace with the technological advancement. Now, digital-based technologies are being unified with marketing performance. In Marketing 4.0 situation, new generation of marketing solutions appeared and took their share in a unique manner. These methods, tools, and techniques appeared as a necessary means in shaping the modern market and acceptable due to globalization. The advancement of this new knowledge has nurtured the development of innovation and competitiveness, which are taking place on a strange scale. On the one hand, governments, enterprises, and ordinary people are heavily

T. Sharma
IMS Unison University, Dehradun, India
e-mail: tripti.sharma@iuu.ac

A. K. Bhatt (✉)
Addis Ababa Science & Technology University, Addis Ababa, Ethiopia

A. Abawa
Addis Ababa University, Addis Ababa, Ethiopia

© The Author(s), under exclusive license to Springer Nature Switzerland AG 2021
S. Awasthi et al. (eds.), *Artificial Intelligence for a Sustainable Industry 4.0*,
https://doi.org/10.1007/978-3-030-77070-9_9

depending on technology, facilitating a greater productivity at work and a refined new standard of living. It is also the means of acquiring new innovative technologies and expertise.

In the present technology-dependent world, public networking became an explicit platform where businesses can expand their marketing promotion campaigns to a large segment of diverse consumers. The approach for communicating with regulars has contorted to a great extent with the coming out of social media; consequently, businesses necessarily need to learn how to use social media in a way that is compatible with their business table (Mangold and Faulds). It is specifically correct for companies that are determined to increase a strong-willed and targeted profit. This review revaluates the present literature that focuses on businesses, market growth, and positioning of social media as an additional room of their marketing strategy.

1.2   Emergence of Marketing 4.0

The marketing strategy is tending or shifting traditionally toward digital marketing to maximize the region and show out the output in the D3I2C model being the secondary objective. This research presents the hypothesis that contemporary technologies are shifting how marketing is separated and that they will alter the recognized concerto of the market, while industries should come to fundamentals with the carryout that having a marketplace share will no longer be passable to lengthen the market leader.

The objective of this primary research is to analyze social media use by the millennial generation and different digital marketing tools influencing customer engagement. Today, many start-up companies are seen launching their business through digital marketing that has greased the process of every business. This made the business an economical, powerful tool of marketing. Today, many start-up companies are seen starting their business through technology mainly in the field of marketing that may be in a product or a service or both. This made the business a reasonable, authoritative tool of marketing. Through a similar objective of expansion, many agencies are grown only with the business plan to reach out to customers and provide assistance to industries with the social media campaign [2, 3, 6].

2   About the Topic

Industry digitization index (IDI) provided by McKinsey is to determine the digital prime life of companies. They recognized three different programs of industries regarding their digital movement, which includes current privileged, industry leaders and the associated circle. Businesses can be separated into ones in which innovations are on an inferior scale, digital migrants enterprises that promptly go after the leader for the change, and organization that alters its business structure in response to adopt the changing technological digitalized climate.

Viewpoints and projection for Marketing 4.0 are the consequences of multifold fluctuations generated by the condition of markets, the incompatible backdrop of global opposition, customers' changing preferences, advent of new technologies, and innovative advertising campaigns. This combination can be regarded as an enormously speedy cybernetic promotion based on an organization of stimuli, advice, and reactions, and which is exceedingly flexible and make distinctive by a thorough sympathetic of the system of a business. Such open and proactive business organizations follow the real-time track of communication and proceedings. The communication from the advertising organization provides a new move toward the organization of marketing chronicles. Dependable development of detailed fundamentals of the system and their associations are desirable to ignite communication. The relations can also be used to exemplify the arrangement where vertices are fundamentals of a digital marketing mix, and the exchanges between the exertion and the customer are to be found along its boundaries. Digitization changes the development in which pay for decisions are finished like counting how the customer looks for in sequence, considers his/her options, and evaluates, relates to the venture, and makes payment for. The procedure after the changes, that is, the one that replaced the traditional pay for decision-making, is known as the "digital consumer path-to-pay for the journey." The changes have been exposed over the past years due to the development of information technologies; marketing is under the process of its transformation and has moved to the point called Marketing 4.0.

3   The Conceptual Model: Design, Direct, Develop, Interventions, Innovation, and Capabilities (D3I2C)

In this segment, we describe the D3I2C conceptual model that was formed as a structure for less flexible organization of marketing proceedings. The D3I2C conceptual model reflects the major thoughts of marketing management in Marketing 4.0 as an autonomous and self-renewing procedure of commencement, altered copy, and expectation of provocation in very self-motivated surroundings.

- **The design stage**: Enterprises are supposed to center on the growth of pertinent indicators that will communicate to the stage of their digital development.
- **The direct stage** is connected to the exactness of the digital marketing approach. Companies in the premature stages of digital development more often don't rely on a customer- and profitability-oriented marketing strategy, whereas enterprises in an advanced or the uppermost stage of digital expansion are listening carefully for the most part on alteration, novelty, and executive.
- **The development** reflects on the talents and management. Firms at the early stages of digital expansion are characterized by a "silo" society. Enterprises at

the middle stage of digital development focus on an integrating culture, while digitally mature companies manifest an integrated and innovative approach.

4 Objectives

Major objectives are as follows:

1. To understand the influence of digital media on millennial generation.
2. To analyze digital marketing tools that influence customer engagement.
3. To study the opportunities of developing a flexible marketing organization.

5 Literature Review

1. **Social Media:** According to Hall and Peszko [7] the term pronounces a new-fangled way in which end users use the World Wide Web and a place where content is uninterruptedly altered by all operators in a sharing and collaborative way.
2. **Innovation Adoption Process (IAP):** The innovation adoption process (IAP) is another instrument that provides all the information related to the consumer's acceptance of new technology. The IAP is the progression through which an individual goes through the innovation–decision process. IAP includes five steps: knowledge of the innovation, forming an attitude toward the innovation, deciding to adopt or reject the innovation, implementation of the innovation, and confirmation of the decision.
3. **Shopper Marketing** is "the planning and execution of all the marketing activities that influence a shopper along, and beyond, the entire path of purchase, from the point at which the motivation to shop first emerges through purchase, consumption, repurchase, and recommendation." [7].
4. **Virtual Brand Community:** The transition from a material economy to an information economy and knowledge-based and interactive communication, together with the enormous development of communication technologies, in particular the rise of the Internet, have led to the growing concept of online brand communities (OBCs), which today can be a great opportunity for business improvement of the firms [1].
5. **Marketing 4.0 and Client Relations Building:** Building relations with customers in the banking sector is changing because of the evolution of information and communications technologies (ICTs) and must be more focused on experiences than on distribution and service quality processes. Service providers need to understand more about their customers than their perceptions of service quality, satisfaction, and loyalty in different distribution channels, such as the Internet and mobile banking [19].

6. **Digital Transformation:** Advanced globalization is meticulously connected with the hastily intensifying technologies producing, dispensing, broadcasting, and consuming information. These technologies are considered as the most important transformation processes of today businesses. They are compelling corporations to bring changes in their traditional marketing approaches, organizational structural reform, and even updating business models [3].

6 Research Methodology

Researcher deployed descriptive research is to satisfy the given objective for which data had been collected from quantitative data. For this, we have constructed the research question for empirical study from the logic of the D3I2C model. The survey was performed based on data collection from Web sources, i.e., Google Form. The link was sent on social media and through email ID.

7 Findings

7.1 Introduction to Questionnaire

1. **Age**

    Figure 1 shows the age in years.

2. **Maximum-Used Medium for the Internet**

    Figure 2 shows the highly used medium for the Internet.

7.2 Millennials' Attitude Toward Social Media

3. **Purpose of Spending Most Time on the Internet** (Fig. 3)

Fig. 1 Age (in years)

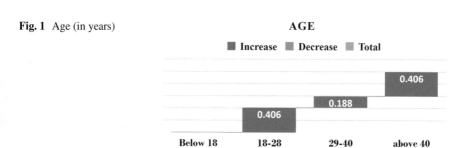

Fig. 2 Highly used
medium for the Internet

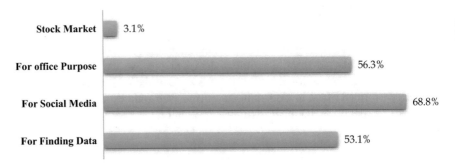

Fig. 3 Purpose of spending time on the Internet

Fig. 4 Time spent on the
Internet

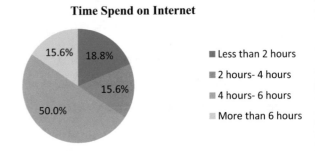

4. **Time Spent on the Internet** (Fig. 4)

5. **Frequency of Social Media (Daily)** (Fig. 5)

6. **Response of Marketing Action** (Fig. 6)

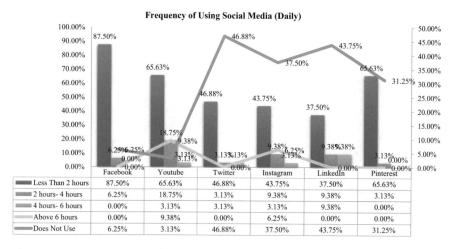

Fig. 5 Frequency of using social media (daily)

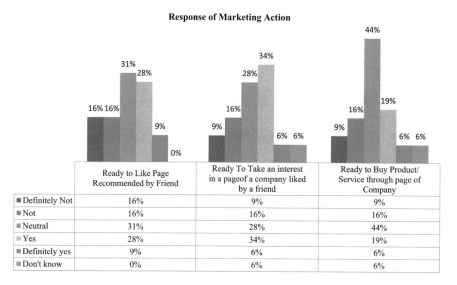

Fig. 6 Response of marketing action

7.3 *Awareness Toward Digital Marketing*

7. **Awareness About Digital Marketing Tool** (Fig. 7)

8. **Information of fast-moving consumer goods (FMCG product) and Real Estate of Different Services** (Fig. 8)

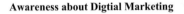

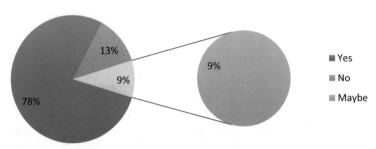

Fig. 7 Awareness about digital marketing

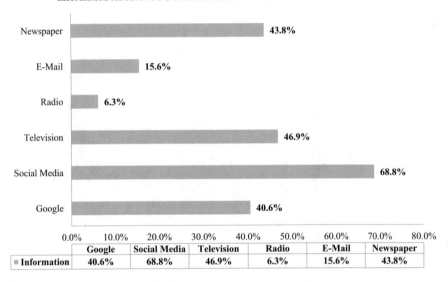

Fig. 8 Information about FMCG and real estate from different media

9. **Maximum-Used Online Media for Getting Information About FMCG** (Fig. 9)

10. **Maximum-Used Online Media for Getting Information About Real Estate** (Fig. 10)

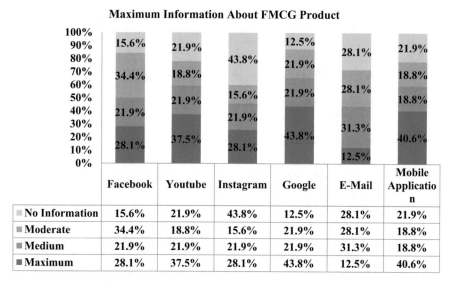

	Facebook	Youtube	Instagram	Google	E-Mail	Mobile Application
▪ No Information	15.6%	21.9%	43.8%	12.5%	28.1%	21.9%
▪ Moderate	34.4%	18.8%	15.6%	21.9%	28.1%	18.8%
▪ Medium	21.9%	21.9%	21.9%	21.9%	31.3%	18.8%
▪ Maximum	28.1%	37.5%	28.1%	43.8%	12.5%	40.6%

Fig. 9 Maximum information about FMCG

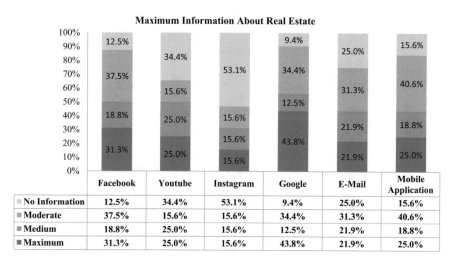

	Facebook	Youtube	Instagram	Google	E-Mail	Mobile Application
▪ No Information	12.5%	34.4%	53.1%	9.4%	25.0%	15.6%
▪ Moderate	37.5%	15.6%	15.6%	34.4%	31.3%	40.6%
▪ Medium	18.8%	25.0%	15.6%	12.5%	21.9%	18.8%
▪ Maximum	31.3%	25.0%	15.6%	43.8%	21.9%	25.0%

Fig. 10 Maximum information about real estate

7.4 Technology on Building Marketing 4.0

11. **What Online Formats Do You Prefer to Learn About Companies and the Products or Services They Offer?** (Fig. 11)

12. **Tone to Learn About Company Products/Services They Offer** (Fig. 12)

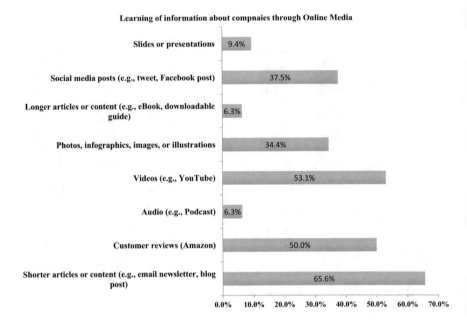

Fig. 11 Learning of information about company

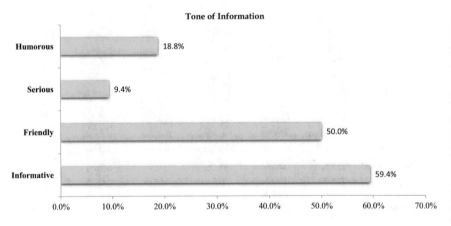

Fig. 12 Tone of information

13. **Media Used for Getting Information About Products/Services They Offer** (Fig. 13)

8   Interpretation

The interpretation of the charts and tables depicts that 59% were millennials and other total responders were above 40 years. The categorization of millennials is ranged from 18 years to 40 years. Also, from the above-given findings, it has been

SHARING COMPANY INFORMATION

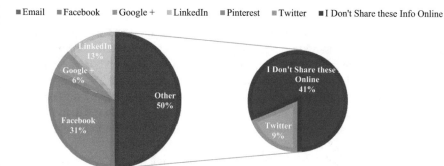

Fig. 13 Sharing company information

found that the source of the Internet that is used foremost is mobile, which means that today society is tending the majority on mobile streaming to find out some source of information and output. With regard to the influence of digital media on millennial generation, it has been perceived that people, mainly millennials, spend majorly 4–6 h on the Internet and reasons behind it are social media, finding data, and official purpose, which state that people are more tended and can be targeted to influence them. It is also found that the least used social media (people do not use daily these platforms) are Twitter (46.3%) and LinkedIn (43.8%) for getting any information, and the most frequently used media are Facebook (87.5%) and YouTube (65.3%).

The study observed the awareness about available digital marketing tools. The majority (78%) are aware of the different tools and their uses. Social media (69%) are the maximum-used, which are used by people to seek any information about real estate and FMCG, and the least used medium is radio (6.3%). It is also perceived that the newspaper (44%) and television (47%) are also the maximum-used media, which means that people also much rely on traditional media in comparison with digital media for seeking information.

Last, the companies also try to publicize the information via digital media, as is seen from past researches that it is the best vital and the cheapest for targeting the population. Audio (6%) and presentation (9%) are the least used for seeking information, and video (53%), along with short articles (blogs) (66%) and customer reviews (50%), is the highest used medium for seeking information. And people also want companies to share much of informative (59%) and also friendly (50%) tone and least serious (9%) informative tone. With the responses, it is seen that customers want companies to share their product and service information once a week (53%) and don't want daily (9%) information.

9   Conclusion

The energetic role of learnedness for business organization and information technologies used for marketing purposes positions substantial hitches for marketing researchers and practitioners. Most of the companies, who wish to go digital in order to be competitive in their consumer segments, are more concentrating on their variables like costs, expenses, and product features, relying largely on their own traditional resources. Businesses are running from the activation mode to the fine-tuning mode uncaring after multinationals working on an international proved information systems and technology-based tools and techniques.

The altering global external milieu is overwhelming organizations to come up with their own innovative systems and promoting business organizations to challenge everything that are chocking them with closed systems. Businesses must investigate all the potential options and discover the appropriate marketing mix, i.e., design stage. In order to establish their goalmouths and advantages for digital transformation, the administration of the company should focus on producing digital strategy and apply marketing mix through technology-based tools and techniques. Business organizations must restructure their businesses into the modern digital business model, i.e., direct stage in consultation of digital experts, and utilize the expertise, information, and experiences kept to complete the accepted artifice of development stage. Marketing experts and administrators must be readily available and easily accessible and should have a proactive approach to answer properly to both market possibilities and the need to advise the precautionary measures for the same. Technology experts are also expected to alarm the business organizations not to be fearful of using the technological tools and interrelated prospects.

References

1. S. Brogi, Online brand communities: A literature review. Procedia Soc. Behav. Sci. **109**, 385–389 (2014)
2. F. Casalo, Guinalíu, Relationship quality, community pro-motion and brand loyalty in virtual communities: Evidence from free software communities. Int. J. Inf. Manag. **30**(4), 357–367 (2010)
3. C. Salkin, M. Oner, A. Ustundag, E. Cevikcan, A conceptual framework for industry 4.0, in *Industry 4.0: Managing the Digital Transformation*, Springer Series in Advanced Manufacturing, (Springer, Cham, 2018), pp. 3–23
4. P. Dallasega, E. Rauch, C. Linder, Industry 4.0 as an enabler of proximity for construction supply chains: A systematic literature review. Comput. Ind. (2018). https://doi.org/10.1016/j.compind.2018.03.039
5. B. Ding, Pharma industry 4.0: Literature review and research opportunities in sustainable pharmaceutical supply chains, in *Process Safety and Environmental Protection*, vol. 119, (Elsevier, 2018), pp. 115–130
6. M. Füller, M. Hoppe, Brand community members as a source of innovation. J. Prod. Innov. Manag. **25**, 608–619 (2008)

7. H. Hall, Peszko, Social media as a relationship marketing tool of modern university. Marketing i Zarządzanie **46**, 41–56 (2016)
8. P. Harris, M. Hendricks, E.A. Logan, A reality check for today's C-Sui Te on industry 4.0, https://assets.kpmg/content/dam/kpmg/xx/pdf/2018/11/a-reality-check-for-todays-c-suite-onindustry-4-0.pdf. Accessed 12 Mar 2019
9. H.Y. Jang, L. Kim, The influence of online brand community characteristics on community commitment and brand loyalty. Int. J. Electron. Commer. **12**(3), 57–80 (2008)
10. M. Laroche, M.R. Habibi, M.O. Richard, To be or not to be in social media: How brand loyalty is affected by social media? Int. J. Inf. Manag. **33**, 76–82 (2013)
11. Y. Liao, L.F.P. Ramos, Past, present and future of Industry 4.0 – A systematic literature review and research agenda proposal. Int. J. Prod. Res. **55**, 3609–3629 (2017)
12. G. Mazurek, K. Nosalska, *The Use of Content Marketing on the B2B Industrial Market – Empirical Research* (Marketing Rynek, 2018), pp. 405–416
13. McKinsey & Company, *Industry 4.0 How to Navigate Digitization of the Manufacturing Sector*, (Mckinsey Digital, 2015). Available at: https://doi.org/10.1080/18811248.1966.9732270
14. O.B. Sezer, E. Dogdu, A.M. Ozbayoglu, Context-aware computing learning, and big data in internet of things: A survey, in *IEEE Internet Things*, (2018), pp. 1–27
15. G.B. Pereira, A. Santos, P.L. de, M.G. Cleto, Industry 4.0: Glitter or gold? A systematic review. Braz. J. Operat. Prod. Manage. **15**(2), 247–253 (2018)
16. B. Schreiber, W.-D. Hoppe, B. Schoenefuss, Future of operations in the digital world (2016). Accessed 13 Mar 2019
17. M.J. Shaw, Web-based e-catalog systems in B2B procurement. Commun. ACM (2000). https://doi.org/10.1145/332833.332845
18. N. Sterev, Marketing leadership: The industry 4.0 need of next generation marketing. Trakia J. Sci. **15**(1), 99–103 (2017)
19. W. Wereda, J. Woźniak, Building relationships with customer 4.0 in the era of marketing 4.0: The case study of innovative enterprises in Poland. Soc. Sci. (2019). https://doi.org/10.3390/socsci8060177

Smart and Intelligent Chatbot Assistance for Future Industry 4.0

Harsh Khatter, Prabhat Singh, Vinay Kumar, and Divya Singh

1 Introduction

Chatbot system is a conversational entity that is using AI algorithms for its implementation and uses it to convert human language into natural language using techniques such as image processing and video processing, natural language processing (NLP), and audio analysis. Chatbot can provide 24×7 customer support experience at a very low maintenance cost. Providing good customer service is one of the major and most important components for any successful business. It is not possible for the business to provide round-the-clock physical customer support to the human user; therefore, a chatbot can be used to enrich the customer support experience so that the customer can get support at any time, at any place, and on any device [1].

This chatbot is built by making use of artificial intelligence algorithms that analyze, inspect, or examine the user's questions (or queries) and understand the user's message what he/she wants to say. We are going to design and develop a system for various stakeholders where users can perform many tasks like food ordering, ticket booking, hospitality, etc. The stakeholders will benefit from their online businesses, e-commerce, education, telecoms, information technology (IT) sectors, healthcare and hospitality, tours and travels, etc. [2, 3]. The proposed model of chatbot will be working across devices such as personal computer (PC), mobile, smart watches, etc.

H. Khatter · P. Singh
Department of Computer Science and Engineering, ABES Engineering College,
Ghaziabad, Uttar Pradesh, India

Affiliated from Dr. APJ Abdul Kalam Technical University, Lucknow, India
e-mail: harsh.khatter@abes.ac.in; prabhat.singh@abes.ac.in

V. Kumar (✉) · D. Singh
Department of Computer Science and Engineering, Amity University Greater Noida Campus,
Greater Noida, Uttar Pradesh, India
e-mail: vkumar@gn.amity.edu; dsingh@gn.amity.edu

© The Author(s), under exclusive license to Springer Nature Switzerland AG 2021
S. Awasthi et al. (eds.), *Artificial Intelligence for a Sustainable Industry 4.0*,
https://doi.org/10.1007/978-3-030-77070-9_10

This chatbot recognizes what users want to say and understands the user's queries and responds to the user simultaneously and answers all the queries. Even if there is any error while writing the query, then chatbot will automatically understand and correct the query and answer according to the situation. The already built-in system of artificial intelligence will understand what the user wants to say and realizes the user's requirements and answers accordingly. So there is no built-in or predefined format for the user to write or ask questions [4, 5].

Chatbots are the machines that have intelligence and will act as machine-to-human conversation systems. The conversation between the human and computer can be done by either text or speech. The natural language processing is used to understand the human text, and speech recognition can be used to understand the speech dialog. In the traditional process of chatbot, the main way to understand the text and reply to the query answer in text form. Chatbots are of two types; i.e., classifications are of two kinds: independent and Web-based chatbots [6, 7].

Independent chatbots are used in stand-alone computers in which there is no need to access the Web, while the Web-based chatbots are using the Web and can be accessed through Web services and can be built on the cloud. In this chatbot system, we are trying to build a Web-based chat application that runs across all devices to provide flexibility and provide real-time conversation that makes the customer service more interactive [8–10].

2   Related Work

Chatbot is built by making use of artificial intelligence algorithms that analyze, inspect, or examine the user's questions (or queries) and understand the user's message what he/she wants to say. The chatbot recognizes what users want to say and understands the user's queries and responds to the user simultaneously and answers all the queries. Even if there is any error while writing the query, then chatbot will automatically understand and correct the query and answer according to the situation. The already built-in system of artificial intelligence will understand what the user wants to say and realizes the user's requirements and answers accordingly. So there is no built-in or predefined format for the user to write or ask questions [11, 12].

The working of chatbot is defined in three stages:

1. **Initial Stage**—In the initial stage, focus on simple Dialogflow–based chatbot where there is no implementation of AI algorithms. The Dialogflow provides us the implementation of natural language processing. NLP will help us to communicate with chatbot using speech by transformation speech into text or vice versa.
2. **Medium Stage**—In this stage, the multiple modules with the implementation of AI algorithms will make the chat application faster and respond to unknown queries searching from search engines also. It will have implementation based on

entities and intents that are used to identify the flow and information from the given query.

3. **Enhanced Stage**—This stage will have the implementation of the interface and application with more furnished AI algorithms so that it can provide support to all kinds of customers. In addition to this, the deployment on the cloud is another phase. It will improve the performance and provide scalability, virtual machines, control, and security [13].

2.1   Product Perspective

When it comes to the emerging market of conversational AI services, it's important to note IBM's traditional leading role in the field. Today, in 2019, all the major tech companies, including Google, Microsoft, Amazon, and even Facebook, as well as many rising start-ups in the field offer natural language understanding and conversational AI tools as a service [14].

These AI services allow simplifying building smart conversational bots. Some platforms provide an end-to-end solution, while others provide components that require some programming in order to compose a bunch of application programming interfaces (APIs) and services into a solution. Here, I'll try to both give an overview and compare the important features of those popular enterprise-grade platforms from the end-user point of view.

All three platforms, Dialogflow, Microsoft's LUIS, and IBM's Watson Assistant, allow you to add examples of utterances and their "intent" category. By training on those examples, it learns to interpret what the user might say into a category that can reflect an action in an app. For instance, a user can say, "I need a cheap flight ticket to Florida"; this can be mapped to an action of a search for a flight ticket [15, 16].

The following subsections describe how the software operates inside various constraints.

2.1.1   System Interfaces

Our chatbot will interact with a DB to store the user's data for improving the user experience. The data from the DB will be used next time when the same user will give the same type of query. Therefore, it improves the performance and efficiency and makes the chatbot working faster.

2.1.2   Interfaces

In our proposed system, we are creating this chatbot by using Google Dialogflow API that provides us the implementation of natural language processing. Using Dialogflow console, we create intents and entities that are used to identify the flow and information from the given query. We need to first create the user interface (UI) for this chatbot by which users can have a wonderful conversation experience. After creating the UI, we start connecting it to the back end using Node.js where we need to implement the services, account handling, conversational experience, and creating intents with webhook (Fig. 1).

2.1.3 Communication Interfaces

It specifies the various interfaces to communications such as wide area network, i.e., Internet. Our chatbot will be able to communicate using the Internet and will respond to each query of the user's using Dialogflow API. This API will identify entities and agents, and it will respond to the user's query on the basis of those entities and agents.

1. If the chatbot is not able to work using an entity and agent, the chatbot will use the Internet and will try to find the answer of the user's query.
2. The chatbot will have only small interactive operations whenever the user wants to perform some operation, and rest of the time, the chatbot will have periods of unattended operations.
3. The chatbot will use Artificial Intelligence Markup Language (AIML) and implementation of algorithms to perform the operations.

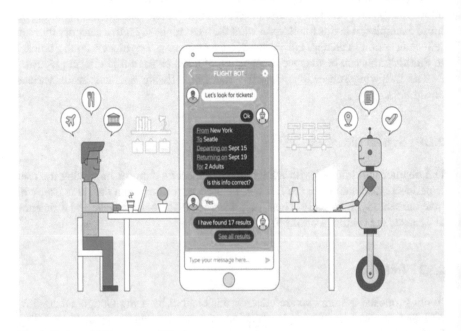

Fig. 1 Real-world interaction of chatbot

4. There is a very less chance that the chatbot will fail as it can use the Internet to respond to the user's query and will try to work properly always.

2.1.4 Site Adaptation Requirements

This chatbot doesn't need any prior software installation at the user or customer site, and any DB will not also be maintained at the customer site. The customer doesn't need to do any modification in his device like desktop, mobile, and smart watch.

The customer will only need to connect to the Internet to use the chatbot so that the chatbot can perform online operations and have a larger efficiency. To have a larger efficiency, the Internet connectivity should be good enough; otherwise, the customer will face issues during chatbot working.

2.2 Product Functions

The chatbot system will be used to provide a great customer service experience among customers of various stakeholders so that they can make their profit and enhance their customer service. This chatbot can be helpful among many sectors such as food industry, travel and tourism, food ordering system, and IT industry and for all enterprises. This chatbot is implemented using Google Dialogflow and deploy over Google Cloud, which will first process the user query and then give an appropriate response to the user's query. It is an intelligence-based system that is used to perform speech-to-text conversion and vice versa. Hence, our chatbot will reduce the human effort and can provide an accurate response to the user's query [17] (Fig. 2).

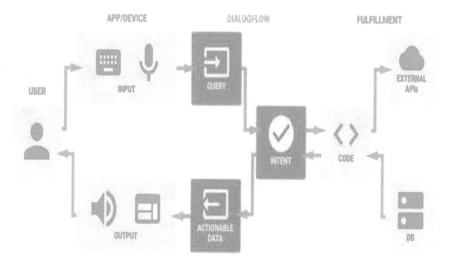

Fig. 2 Product function of chatbot

2.3   Assumptions and Dependencies

The availability of Google Cloud Platform (GCP) for deploying the chatbot can change the software requirements specification (SRS). It will affect the security, virtual machines, and the performance of chatbot. It can also affect the availability of chatbot on all devices like smart watch. It may be difficult to provide availability of chatbot on smart watch without deploying on Google Cloud Platform. The availability of Google Cloud Platform includes some parameters like time availability, cost of platform, and working of chatbot on GCP [18, 19].

3   Architectural Design

The concept is the idea that forms the backbone of the entire project. Diagrams help architects to show that concept visually and clearly, by removing all of the other little details that can make the concept harder to see (Fig. 3).

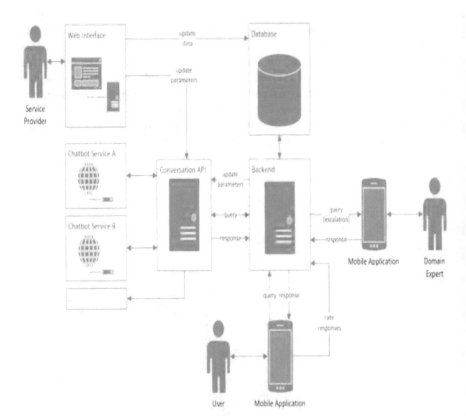

Fig. 3 Functional architecture of chatbot

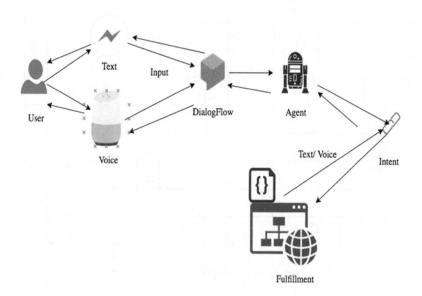

Fig. 4 System workflow with Dialogflow API

Here is the basic and brief view of the architectural view of system design. It shows the system workflow with the Google Dialogflow API for the selection of responses and the text/voice-based output to the user [20] (Fig. 4).

Now here is the block view of the system design that shows how the user is sending the query to the chatbot and how this query is reaching to the Dialogflow. With this, there is also the intent matching and the fetching of the response from the external APIs and databases (Fig. 5).

So till now, we saw the architectural view of our system. Now, we will look at the further system requirements and other diagrams.

But before that, here is the high-level architecture of the chatbot with the concise and effective representation (Fig. 6).

As you can see that there are three components present named as Google Dialogflow, Application DB, and Middleware that are the important parts required for the working of chatbot.

Therefore, this is all about architecture diagrams of the required chatbot and its working. Now look for further class diagrams and all.

3.1 Class Diagrams

Now we can see the class diagrams for the different agents available in our chatbot. As we know that this chatbot is multimodule-based, we need not to mention all the class diagrams here. Therefore, here are the few examples of class diagrams of a few

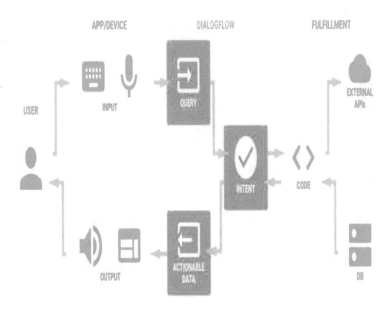

Fig. 5 Block view of the system design

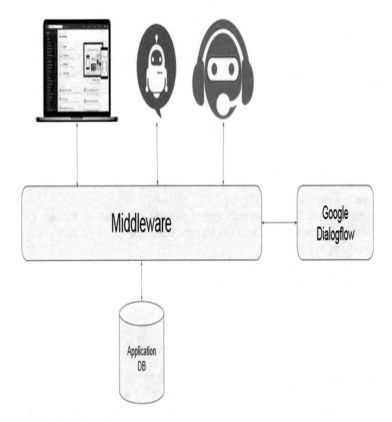

Fig. 6 High-level architecture

agents such as flight booking, pizza ordering, hotel booking, movie searching, and so on [15, 21].

3.2 Pizza Ordering Agent (Class Diagram)

In this agent, the customer can order the pizza with the help of this chatbot also. There are many classes available as mentioned above in the diagram (Fig. 7).
     Happy conversation is to flow as:

- Bot: Hi "Customer Name," welcome to Dominos!
- Bot: How can I help you?
- Customer: I would like to order a pizza.
- Bot: Great, what pizza would you like to try today?
- Customer: What all options do you have?
- Bot: Would you like to go for vegetarian or nonvegetarian today?

Pizza Ordering Agent (Class Diagram):

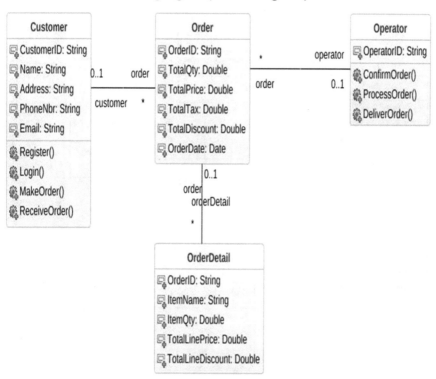

Fig. 7 Chatbot of pizza ordering agent class diagram

- Customer: Vegetarian/nonvegetarian.
- Bot: Display the options in the choice selection.
- Customer: I would like to order "XYZ."
- Bot: Great choice, what size would you like to order?
- Customer: I would like to order a medium size.
- Bot: Please let me know the address to deliver the pizza.
- Customer: "Address".
- Bot: Please let me know what time would you like to enjoy your pizza?
- Customer: In about an hour.
- Bot: (Display the entire order.) Thank you!

3.3 Hotel Booking Agent (Class Diagram)

Let's check the happy conversation with the hotel booking chatbot, and how it will respond to the user queries and help the customer to book hotels for them. Here you can experience a new way of customer service with full satisfaction and ease of need (Fig. 8).

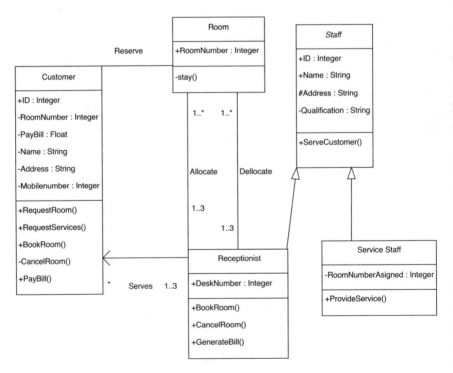

Fig. 8 Class diagram for hotel management chatbot

Welcome to StayBot, an all-in-one communication platform for hotels that replies to FAQs of guests and incentivizes them to book.

Click in "View Details" to know more about each offer and book your stay with us.

3.4 Data Flow Diagram

Now, we can have a look over the data flow diagrams (DFDs) for any of the above-mentioned agents. Let's take the example of a flight booking agent and start with its data flow diagrams.

Data Flow Diagram for Login In this DFD, the representation of the login as a user/an admin for the flight booking process is shown here. On the basis of the role of the user, there will be different options available for further processing of the system functionalities (Fig. 9).

Here is the representation of the DFD for the login functionality, and in this way, the login will work for any of the agents provided in chatbot. After the login, more functionalities exist for customer and admin separately whose DFDs are shown below.

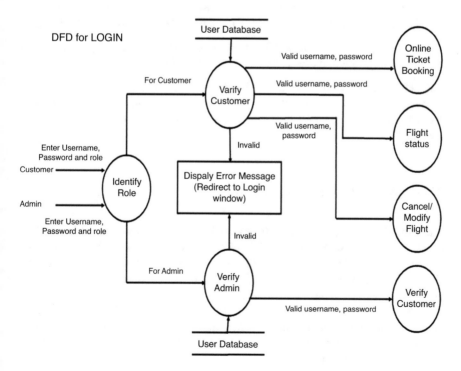

Fig. 9 Data flow diagram of chatbot

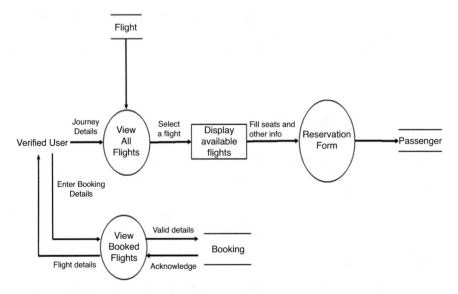

Fig. 10 Data flow diagram of online booking chatbot

Data Flow Diagram for Online Booking In this DFD, there is the representation of the data flow during the process when a user wants to book their flight ticket. Here is the complete description flow of the online booking process [11] (Fig. 10).

In this DFD, we can see how it first verified the user and input the journey details, and in this way, only the user is able to make a booking of the flight.

Data Flow Diagram for Post Ticket Booking (Fig. 11)
In the above DFD, we can see the flow of more functionalities provided for the user and how he/she can manage his/her bookings and cancelled ticket or update their bookings.

4   Implementation and Results

The implementation part started where all the technical works are there like coding, testing, maintenance, and so on. To look at the implementation part with all the required syntax and libraries used during the execution of the application, here is the close look at the architectural view of the working of the chatbot with the integration of Web apps on different platforms (Fig. 12).

From here, we will start looking for the implementation of the application as brief snapshots of implementation phases. Before that, we need to understand the required software specifications as mentioned above because the implementation can be performed on this software only.

Initialization of Implementation Phases
• Working on rule-based chatbot.

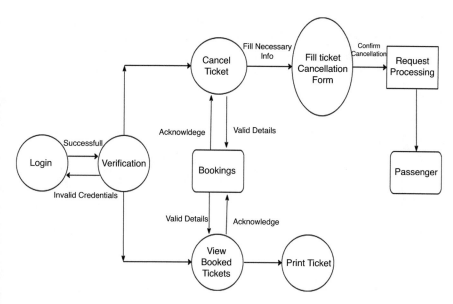

Fig. 11 Data flow diagram of post ticket booking chatbot

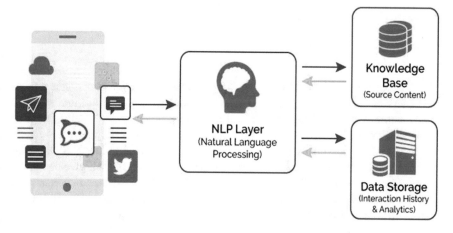

Fig. 12 Chatbot integrated with Web apps

- Created agent.
- Created intents and entities.
- Worked upon small talks feature of Google Dialogflow.
- Worked on single module–based chatbot (flight booking module).
- Implement webhook at Dialogflow.
- Implement Web demo of testing agent.
- Integration with slack (Fig. 13).

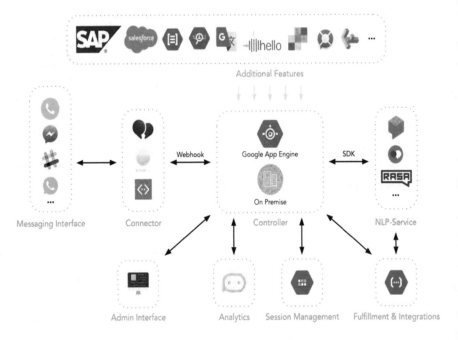

Fig. 13 Results of chatbot on apps

4.1   Results

This chatbot system is used to provide a great customer service experience among customers of various stakeholders so that they can make their profit and enhance their customer service. This chatbot can be helpful among many sectors such as food industry, travel and tourism, food ordering system, and IT industry and for all enterprises. This chatbot is implemented using Google Dialogflow and deployed over Google Cloud, which will first process the user query and then give an appropriate response to the user's query. It is an intelligence-based system that is used to perform speech-to-text conversion and vice versa. Hence, our chatbot will reduce human effort and can provide an accurate response to user's query.

5   Conclusion

Chatbots are fully-functional and autonomous systems that are used mostly to solve the queries of customers. The future scope of this chatbot is to include the profitable for big enterprises as well as to provide many benefits for the business and larger enterprises. As with the help of AI-based chatbots, any enterprise can provide very rich conversational experience during customer support.

Therefore, AI-based chatbot helps many organizations and enterprises to make their customer support automatic, and the conversation experience will be more interesting just like with human; they can provide their service for 24 × 7.

Google Dialogflow seems the easiest to integrate and put into action. Dialogflow has the most compatible conversational platforms among the pack. Each platform can make use of the Google Dialogflow agent with the click of one button. Much like Amazon Lex, Google Dialogflow can also create a webhook request to fulfill our projector rentals. This fulfillment process can be integrated into any of the supported platforms. Google's Dialogflow provides out-of-the-box connection to messaging and voice UI channels such as Facebook Messenger, Slack, and Google Assistant. Microsoft also provides on top of their Azure Cloud service an out-of-the-box connection to Skype and Skype for Business in addition to Facebook Messenger and other typical chatbots and messaging channels.

IBM's Watson Assistant provides a built-in connection to Facebook Messenger, as well as Slack, WordPress, and custom website chat. When it comes to letting your bot answer questions from your database or perform tasks such as booking a flight ticket, it's very much up to you to decide where to host your back-end code. Dialogflow that is backed by Google's acquired Firebase lets you run "serverless" custom Javascript code on their cloud. Microsoft Azure provides their own version of Node.js app that runs in the cloud for developing custom bots.

References

1. D. Sorna Shanthi, S. Keerthana, P.K. Nandha Kumar, D. Nithya, Hexabot: A text-based assistive chatbot to explore library resources. (IJEAT) ISSN: 2249-8958 **8**(3S) (2019, Feb)
2. A. Atiyah, S. Jusoh, S. Almajali, An efficient search for context-based chatbots, in *2018 8th International Conference on Computer Science and Information Technology*, (2018)
3. G. Hiremath, A. Hajare, P. Bhosale, R. Nanaware, K.S. Wagh, Chatbot for education system. Int. J. Adv. Res. Ideas Innov. Technol. **4**(3) (2018)
4. K. Bala, M. Kumar, S. Hulawale, S. Pandita, Chat-bot for college management system using A.I. IRJET **04**(11) (2017, Nov)
5. S. Wailthare, T. Gaikwad, K. Khadse, P. Dubey, Artificial intelligence based chat-bot. IRJET **05**(03) (2018, Mar)
6. H. Khatter, A.K. Ahlawat, An intelligent personalized web blog searching technique using fuzzy-based feedback recurrent neural network. Soft. Comput. **24**(12), 9321–9333 (2020). https://doi.org/10.1007/s00500-020-04891-y
7. A. Patil, K. Marimuthu, A. Nagaraja Rao, R. Niranchana, Comparative study of cloud platforms to develop a chatbot. Int. J. Eng. Technol. **6**(3), 57 (2017)
8. S. Chawla, Introduction to natural language processing (2018, May 29). [Online] Available: https://medium.com/greyatom/introduction-to-natural-language-processing-78baac3c602b
9. J. Brownlee, A gentle introduction to Bag-of-words model (2017, Oct 9). Processing. [Online] Available: https://machinelearningmastery.com/gentle-introduction-bag-words-model/
10. M. Dhahiya, A tool of conversation: Chatbot. Int. J. Comput. Sci. Eng. **5**(5), 158–161 (2017, May)
11. K.V. Wadanka et al., Chatbot : An application of AI. Int. J. Res. Eng. Sci. Manag. **1**(8), 139–141 (2018, Sep)

12. V. Aggarwal, A. Jain, H. Khatter, K. Gupta, Evolution of chatbots for smart assistance. Int. J. Innov. Technol. Explor. Eng. **8**, 77–83 (2019)
13. R. Pascanu et al., On the difficulty of training recurrent neural networks, in *International Conference on Machine Learning*, (Atlanta, Georgia, USA, 2013)
14. H. Khatter, B.M. Kalra, A new approach to Blog information searching and curating, in *Proceedings to Sixth International Conference on Software Engineering CONSEG*, (Indore, India, 2012), pp. 1–6
15. H. Khatter, A.K. Ahlawat, Analysis of content curation algorithms on personalized web searching, in *Proceedings of the International Conference on Innovative Computing & Communications (ICICC)*, (New Delhi, 2020), pp. 1–4. https://doi.org/10.2139/ssrn.3563374
16. H. Khatter, M.C. Trivedi, B.M. Kalra, An implementation of intelligent searching and curating technique on Blog Web 2.0 tool. Int. J. u e-Serv. Sci. Technol. **8**(6), 45–54 (2015)
17. Elvis, Deep learning for NLP: An overview of recent trends (2018, Aug 24). [Online] Available. https://medium.com/dair-ai/deep-learning-for-nlp-an-overview-of-recent-trends-d0d8f40a776d
18. B. Galitsky, A content management system for chatbots, in *Developing Enterprise Chatbots*, vol. 1, (Springer, 2019, Apr), pp. 253–326
19. N. Ali, M. Hindi, R.V. Yampolskiy, Evaluation of authorship attribution software on a chatbot corpus, in *International Symposium on Information, Communication and Automation Technologies*, (2011, Oct)
20. E. Go, S.S. Sundar, Humanizing chatbots: The effects of visual, identity and conversational cues on humanness perceptions. Comput. Hum. Behav. **97**, 304–316 (2019)
21. P. Singh, H. Khatter, S. Kumar, Evolution of software-defined networking foundations, in *Evolution of Software-Defined Networking Foundations for IoT and 5G Mobile Networks*, vol. 1, (2021), pp. 98–112

Analyzing Subspace Clustering Approaches for High Dimensional Data

Parul Agarwal and Shikha Mehta

1 Introduction

A huge body of knowledge can be exploited through clustering techniques. Clustering algorithms make use of knowledge present in input data. It uncovers the hidden patterns existing in data. Clustering is one of the effective data mining techniques for grouping similar data objects [1]. The grouping is made on the basis of attributes. With the tremendous growth in the technology of information sciences, huge data is being produced. Such data are sensor data, web data, bioinformatics data and many more. The large volume of data not only includes a large number of rows but also a large number of attributes. The datasets with many columns/features are termed as high dimensional data. Clustering high dimensional data poses a number of computational challenges [2]. Traditional clustering algorithms' efficacy degrades on high dimensional data. This problem is termed as "curse of dimensionality." This is because traditional clustering algorithms find clusters in all dimension space. However, clusters might exist in a few subsets of dimensions. All dimensions might not be important for all clusters. Considering all dimensions for clustering might hide the relevant dimensions. Another challenge is the distance measure. Distance between data objects becomes meaningless in high dimensional data [3]. All objects appear equidistant from each other. Most of the traditional clustering algorithms use distance measure for grouping data objects. New techniques are developed to find relevant dimensions and overcome challenges of high dimensional clustering.

Relevant dimensions are selected through feature selection methods and, subsequently, selected features to participate in complete clustering method. However,

P. Agarwal (✉) · S. Mehta
Department of Computer Science and Information Technology, Jaypee Institute
of Information Technology, Noida, India
e-mail: parul.agarwal@jiit.ac.in; shikha.mehta@jiit.ac.in

© The Author(s), under exclusive license to Springer Nature Switzerland AG 2021
S. Awasthi et al. (eds.), *Artificial Intelligence for a Sustainable Industry 4.0*,
https://doi.org/10.1007/978-3-030-77070-9_11

clusters might exist in different subspaces. Dimensions participating in one cluster might be irrelevant for another cluster. Hence to find clusters in different subspaces as shown in Fig. 1 (each circle represents one cluster), subspace clustering methods are employed [4]. It is an extended version of traditional clustering algorithms. Subspace clustering algorithms confine the search in a certain manner, i.e., either top-down or bottom-up approach so that it is able to determine clusters existing in various subspaces. Further, subspace clusters can be non-overlapping or overlapping in dimensions/objects.

The significance of subspace clustering is intensifying from last two decades. Figure 2 shows the percentage of papers published on various methods of high

Fig. 1 Representation of clusters in subspaces

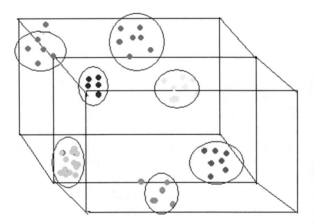

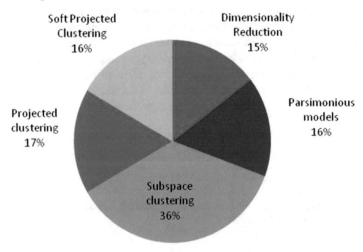

Fig. 2 Papers published on different techniques of high dimensional clustering

dimensional clustering from the year 2000 onwards. The main techniques for handling high dimensional clustering [5, 6] are as follows:

(a) Dimensionality reduction
(b) Parsimonious models
(c) Projected clustering
(d) Soft projected clustering
(e) Subspace clustering

It was observed that only 15% of work is done in high dimensional clustering through dimensionality reduction approach. Sixteen percent of papers are published for clustering high dimensional data using soft projected clustering and parsimonious models. Seventeen percent of the work is dedicated to projected clustering. The large amount of work for high dimensional clustering is being done by subspace clustering approach. This proves that subspace clustering methods are gaining considerable attention in current research.

This study is entailed to answer the critical research questions identified below.

 (i) What are the major challenges faced by traditional clustering algorithms to cluster high dimensional data?
 (ii) What search techniques are being used in subspace clustering to determine subspaces?
(iii) What are the evaluation measures for comparing subspace clustering algorithms?
(iv) What is the current scenario of subspace clustering?
 (v) What are the research gaps in the literature and the future prospects of subspace clustering?

This chapter presents a combined review of approximately all subspace clustering algorithms belonging to different classes. Various evaluation measures required for comparing the clusters and subspaces are also presented. Additionally, statistical data of subspace clustering approaches published in different years and different repositories are provided. This chapter is targeted to researchers planning to work in subspace clustering area. It provides a roadmap of research in subspace clustering approach for high dimensional data. The chapter also presents the application areas, identifies gaps in present work, and suggests future opportunities for research in this field.

The chapter is divided into following subsections: Sect. 2 presents challenges in subspace clustering, Sect. 3 gives various classifications of subspace clustering approaches, Sect. 4 presents evaluation measure for comparing subspace clustering algorithms, and Sects. 5 and 6 depict the literature survey and empirical assessment of subspace clustering algorithms, respectively. Applications and future prospects are illustrated in Sect. 7, and Sect. 8 finally concludes the chapter.

2 Challenges in Subspace Clustering

The major goal of clustering is not only to find similar groups of data points but to find high-quality groups within a reasonable time. In cases where clusters exist in different subsets of dimensions, it is essential for clustering algorithms to determine effective clusters along with relevant dimensions. Subspace clustering algorithms have proved to be efficient for extracting clusters from high dimensional datasets [7]. However, a number of challenges persist which needs to be focused before developing an efficient subspace clustering algorithm. These challenges are as follows:

(a) It is hard to determine the subsets of dimensions where data points are similar. This is because a number of dimensions are large and possible combinations are huge.
(b) It is quite difficult to determine the distribution of data within subspaces. If data is near the cluster center and far from another subspace center, then clustering is easy; otherwise it is difficult to cluster within subspaces.
(c) There may be overlapping subspaces that mean few dimensions may be common in few subspaces. Clustering becomes even more complicated in case of overlapping subspaces.
(d) There are possibilities of noisy data in the dataset. That means some data points might not belong to any subspace or if they are part of any subspace, they are not part of a cluster of particular subspace. Handling of such data becomes a challenge for subspace clustering.
(e) It is difficult to understand which clustering algorithm and subspace strategy are appropriate for a given problem. As the number of subspaces and the dimension of each subspace is unknown, the problem becomes intricate.

Due to the above challenges encountered in subspace clustering approaches, there has been a scope of improvement in these algorithms. Subspace clustering not only determines the clusters in the dataset but also the subspaces in which these clusters are present. The next section presents the review of subspace clustering classification.

3 Classification of Subspace Clustering Approaches

In order to determine a group of similar data points in different subspaces, subspace clustering algorithms are employed. This section discusses the various categories of subspace clustering approaches on the basis of different parameters. Figure 3 depicts the categories of subspace clustering algorithms on the basis of parameters required in algorithms for forming clusters from data objects [8]. These are as follows:

Cell-based Subspace Clustering: This approach is also called a grid-based approach. It makes use of an approximate number of cells required to form a cluster.

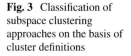

Fig. 3 Classification of subspace clustering approaches on the basis of cluster definitions

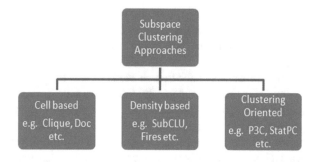

The cluster description starts with a minimum width "w" of a number of cells. Each cell contains a minimum threshold number of objects. The cells of a cluster are either of fixed grid size or variable in number forming hypercube of width "w." The cells participating in clustering uses subsets of dimensions of datasets. Hence the relevant dimensions to a particular cluster are determined. Irrelevant dimensions, not participating in clustering, expand on other cells. Few cell-based algorithms are CLIQUE, DOC, MINECLUS, SCHISM, etc.

Density-based Subspace Clustering: This approach is able to determine the clusters of arbitrary shapes. It is dependent upon the density of data objects lying in datasets. This approach separates the dense region from sparse region. Density is determined through distance measure. The parameters used in algorithms are the least number of points "minpts" required to form a cluster and "epsilon" distance among the neighboring points. The dense region is formed by counting the number of points "minpts" within "epsilon" neighboring distance. Any region not satisfying the "minpts" and "epsilon" properties is not able to form a cluster and is termed as sparse region. Some density-based subspace clustering algorithms are FIRES, INCY, SUBCLU, etc.

Clustering-Oriented-Based Subspace Clustering: This approach does not provide any requirements for cluster formation. It is not dependent on any cluster definition or on input parameters to form a cluster. As the name suggests it gives statistical orientation properties of total clusters formed. It means it defines properties of resultant clusters formed like a number of clusters formed, average dimensionality per cluster, etc. Clustering oriented approach is more suitable to datasets of varied distributions. Some algorithms of this approach are STATPC, P3C, PROCLUS, etc.

Subspace clustering approaches confine their search in such a manner that clusters existing in different subspaces are extracted [3]. The search either proceeds from single-dimensional to full dimensional dataset (bottom-up) or full dimensional to single dimensional dataset (top-down). Each search method is defined as follows:

Bottom-Up Subspace Search Method: It is a grid-based method which starts from single dimension. This method follows an a priori approach [1] to determine relevant dimensions. The method begins with forming similar groups in single dimensions on the basis of density threshold parameters and grid size. The data objects participating in single dimension will also participate in multi-dimensional grids. This approach detects the noise and also determines overlapping subspace clusters. However, it may find redundant subspaces or clusters across the dataset. Some algorithms of bottom-up approach are CLIQUE, ENCLUS, DOC, etc. (Fig. 4).

Top-Down Subspace Search Method: It is an iterative method which starts from entire dimensions of dataset. Initially, each dimension is assigned equal weights and clustering begins. After clustering, the weights of each dimension for each cluster are updated. In the next iteration, updated weights are used, clustering proceeds, and again weights are updated. Dimensions with highest weights for a cluster are relevant dimensions. This is expensive method which requires many iterations. Input parameters used in this approach are number of clusters and size of subspaces. Both the parameters are difficult to decide before clustering. This method finds non-overlapping clusters. Some algorithms are PROCLUS, FINDIT, MAFIA, etc.

Data objects of clusters determined from subspace clustering approaches may or may not be aligned along the axis [9, 10]. On this basis, subspace clustering approaches are divided into two categories as shown in Fig. 5.

Axis Aligned: Clusters determined along the parallel axis to data space are axis-aligned. These subspace clusters could be determined with low computational complexity. Number of subspaces determined from this approach are fixed in number, e.g., CLIQUE and DOC.

Non-axis Aligned: the subspace clusters determined in the arbitrary orientation of data space are non-axis aligned. Clusters may be expressed in a better way using

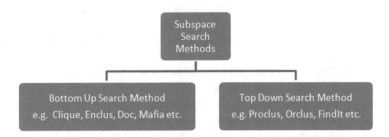

Fig. 4 Classification of subspace clustering approaches on basis of search methods

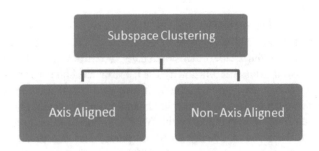

Fig. 5 Classification of subspace clustering approaches on basis of axis alignment

this approach. However, computation complexity is quite high in finding clusters in arbitrary orientation. Subspaces may go infinite in number, e.g., Orclus.

4 Evaluation Measures for Subspace Clustering Algorithms

This section describes the various systematic evaluation measures used for comparing objects and subspaces of clusters formed. However, there are no standard criteria defined for comparing the subspaces or clusters formed from subspace clustering approach. In literature, researchers have employed different evaluation measures for performance assessment of subspace clustering algorithms. A common ground of comparing subspace clustering algorithms is lacking. This is because true cluster labels along with relevant dimensions are lacking in datasets. The paper presents thorough evaluation measures shown in Fig. 6, required for comparing subspace clustering algorithms [8, 9].

Object-Based Validation Measures: This type of validation considers the data objects participating in the clustering process. The various measures are as follows:

F1_Measure – It is the harmonic mean of precision and recall values. This measure ensures that actual cluster (found cluster from the algorithm) should mask maximum objects of true cluster (already given class in dataset) and unmask the objects of different clusters. This is computed from the confusion matrix [1]. Let TP is true positive, i.e., objects of actual cluster are same of true cluster; TN is true negative, i.e., objects of different true clusters are mapped to different actual

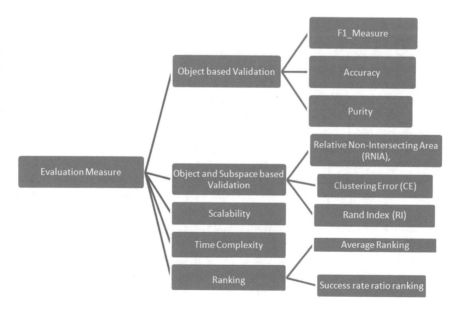

Fig. 6 Evaluation measures for subspace clustering approaches

clusters; FN false negative, i.e., objects of same true cluster belong to different actual cluster; FP false positive, i.e., objects of different true cluster belong to same actual cluster. Precision is minimum mapping of objects from other clusters, while recall is maximum mapping of objects from same true cluster. It is given by Eqs. 1, 2, and 3:

$$\text{Precision} = \frac{TP}{TP + FP} \tag{1}$$

$$\text{Recall} = \frac{TP}{TP + FN} \tag{2}$$

$$F1\_\text{Measure} = \frac{2 * \text{Precision} * \text{Recall}}{\text{Precision} + \text{Recall}} \tag{3}$$

Accuracy: Accuracy is the ratio of a number of objects of the actual cluster correctly mapped with objects of true cluster by total objects. It can be calculated using confusion matrix through following Eq. 4:

$$\text{Accuracy} = \frac{TP + TN}{TP + TN + FP + FN} \tag{4}$$

Purity: This measures the purity or homogeneity of actual clusters determined from clustering algorithm with respect to true clusters. It can be calculated as follows:

$$\text{Purity} = \frac{TP}{TP + TN + FP + FN} \tag{5}$$

Object and Subspace-Based Validation Measures: This type of validation considers not only the data objects but the relevant subsets of dimensions participating in the clustering process. These measures actually evaluate how well clusters are formed in various subspaces. In these measures, an object of a dataset is assumed to be divided into number of sub-objects which spans across all the dimensions of datasets. Hence a subspace cluster consists of relevant dimensions along with sub-objects. An object shared between the subspace clusters with disjoint dimensions will have different sub-objects. Thus, subspace clusters having same objects with different relevant dimensions do not overlap. The various measures are as follows:

Relative Non-Intersecting Area (RNIA): This measure ensures the maximum number of sub-objects found in actual subspace cluster maps the true subspace cluster. A confusion matrix is formed from actual subspace clusters versus true subspace clusters [9]. The matching sub-objects with each subspace clusters are filled in the matrix. Let "U" is the total number of sub-objects participating in true or actual subspace clusters. Same sub-objects participating found in both true and actual cluster (with the same dimension) is counted once. Let "I" is the intersecting subjects of true and actual subspace clusters, or it is the sum of all elements of confusion matrix. Then RNIA is calculated as follows:

$$RNIA = \frac{U - I}{U} \tag{6}$$

Clustering Error (CE): RNIA gives the same result in a case when several actual clusters match one true cluster or one actual cluster matches true cluster. Clustering error maps one actual cluster to almost one true cluster and vice versa. CE is also calculated from confusion matrix formed by above method. Let D is the sum of all principal diagonal elements, then CE is given as:

$$CE = \frac{U - D}{U} \tag{7}$$

Rand Index: This index is measured on the basis of counting the pair of sub-objects that do or do not participate in clustering. All sub-objects that do not overlap are considered as single clusters. A confusion matrix is created considering all sub-space clusters including singleton clusters. There are four labels important, i.e., N11, pair of sub-objects that are same in both actual and true subspace clusters; N10, pair of sub-objects that are same in a true cluster but not in actual cluster; N01, pair of sub-objects that are same in actual cluster but not in true cluster; and N00, pair of sub-objects that are different in both actual and true subspace clusters.

$$1 - Rand_{index} = \frac{N10 + N01}{N} \tag{8}$$

where $N = \frac{U * (U - 1)}{2}$, U is the union of all sub-objects participating in clustering.

Scalability: Scalability is the measure which could be used to visualize and analyze the behavior of any algorithm. In subspace clustering algorithms, scalability is measured in terms of dimensionality. It depicts the performance of an algorithm with the increase in a number of dimensions of dataset. The graphs are plotted which could be shown for any evaluation measure with respect to dimensions. The X axis shows the dimensions and Y axis depicts evaluation metric like F1_Measure or accuracy etc. Scalability of subspace clustering algorithms is shown in the latter part of the chapter.

Time Complexity: Run time complexity measures the time taken by an algorithm to cluster a given dataset. It is represented in the form of graphs to compare subspace clustering algorithm with respect to dimensions. Few examples of representation are shown in [8].

Ranking: Subspace clustering algorithms can be ranked on the basis of average ranking (AR) and success rate ratio ranking (SRR) [11]. The average rank of an algorithm is computed by taking the mean of ranks on all dataset on basis of any evaluation measure. Let r_j^i be the jth algorithm rank for i^{th} dataset. The average rank of each algorithm on total "n" datasets is computed using following Eq. 9:

$$r_j = \frac{\sum_{i=1}^{n} r_j^i}{n} \tag{9}$$

Success rate ratio rank (SRR) computes the ratio of success rates in a pair of algorithms. This method is useful in determining the significant differences in algorithms. In SRR ranking, an algorithm and dataset are taken and accuracy (any evaluation measure) ratio is calculated with respect to other algorithms. This ratio is computed by the following Equation:

$$SRR_{j,k,j\neq k}^{i} = \frac{acc_j^i}{acc_k^i} \tag{10}$$

where "i" is the dataset, "j" is the algorithm for which SR is calculated, and "k" is the compared algorithm different from j. Hence SRR is calculated for algorithm "j" with reference to algorithm "k" on an ith dataset. Likewise, with the same pair of algorithms, SRR is calculated on all datasets. Subsequently, the mean of SRR is computed for all n dataset using Eq. 11.

$$SRR_{j,k,j\neq k} = \frac{\sum_{i=1}^{n} SRR_{j,k,j\neq k}^{i}}{n} \tag{11}$$

In the same way, the algorithm "j" is paired with all other algorithms on all datasets and overall SRR for algorithm "j" is given by:

$$SRR_j = \frac{\sum_{k} SRR_{j,k,j\neq k}}{m-1} \tag{12}$$

where "m" is a number of subspace algorithms used for comparison. Similarly, SRR for all algorithms against each and every algorithm is calculated and ranked.

The further sections discuss the literature survey outline with analysis of subspace clustering algorithms on a few evaluation metrics.

5   Literature

There is a number of good surveys made by researchers on high dimensional clustering but very few are on subspace clustering. Assent [2] made a brief survey of high dimensional clustering on different types of datasets. Authors have also shown different methods adopted for clustering high dimensional datasets. Clustering in high dimensions is proposed through various models by [6]. A survey on models of high dimensional clustering datasets is given by [12]. Kriegel et al. [5] presented a survey on clustering high dimension data through subspace, correlation, and pattern-based clustering methods. Steinbach et al. illustrated the challenges of high dimensional clustering in detail and proposed a concept-based model for handling dataset with large attributes [13]. Fahad et al. made a survey describing various classes of clustering algorithms along with different evaluation metrics [14]. A comparative analysis of different swarm intelligence-based clustering algorithm is depicted in [15]. A new clustering algorithm with modified flower pollination algorithm is shown in [16]. However, the various studies discussed above mainly focused on clustering high dimensional datasets; limited studies have been performed on subspace clustering approaches. Parson [3] made a comprehensive survey on subspace clustering approaches portraying different subspace search methods. Müller [8] depicted the different evaluation measures as well as different categories of subspace clustering algorithms. The work [17] also provided a WEKA platform for evaluating various subspace clustering algorithms. Subspace clustering through evolutionary technique is proposed by [18]. Table 1 presents the review on various subspace clustering algorithms along with evaluation metrics used, type of dataset applied, and maximum dimensions evaluated and drawbacks.

It can be observed from Table 1 that maximum dimension evaluated in high dimensional dataset is 5920, but 6144 is also there. Many algorithms have been proposed on subspace clustering to handle high dimensional datasets. A Monte Carlo-based subspace clustering algorithm [44] is proposed by Olson et al. and evaluated against subspace and projected clustering algorithms on real and synthetic datasets. The significance of subspace clustering approaches is increasing in this big data era. Figure 7 depicts the number of papers published in a span of 5 years (shown on the x-axis).

Subspace clustering approaches also gained importance in various repositories of publications. Figure 8 shows the percentage of subspace clustering papers deposited in different repositories.

Figure 9 gives the percentage of papers published in conference or journals. 53% of subspace clustering techniques was published in conferences and 47% of papers were published in journals.

The next section illustrates the empirical comparison of subspace clustering algorithms on real and artificial datasets.

6   Empirical Assessment

Subspace clustering algorithms are mainly classified as cell, density, and clustering-oriented-based algorithms [8]. The algorithms belonging to these categories cover almost all subspace clustering algorithms. Present section shows the empirical assessment of subspace algorithms on the basis of two evaluation measure i.e. F1_Measure and accuracy. The algorithms are compared on the basis of ranking and scalability.

(i) Ranking of Subspace Clustering Algorithms: Rank of algorithms [11] on real and synthetic datasets are made independently on accuracy and F1_Measure.

Table 1 Survey of various subspace clustering algorithms

S. no.	Year	Authors and references	Approach used/ proposed	Evaluation measure	Type of dataset	Maximum dimensions evaluated	Gaps
1.	2019	Pan Ji et al. [19]	Adaptive low-rank kernel subspace clustering	Clustering error	Motion and face dataset	2016	Need to manually select kernel function
2.	2019	Jufeng Yang et al. [20]	Subspace clustering via good neighbors	Accuracy and NMI	Image real dataset	2016	Difficulty in finding good neighbors
3.	2018	Paul and Nayagam [21]	PROCLUS	Accuracy	Student performance studies	–	Authors could have used some latest algorithms to analyze the performance of students
4.	2018	Manolis C. Tsakiris Rene Vidal [22]	Sparse subspace clustering	–	–	–	The paper gives theoretical proves without implementation
5.	2018	Canyi Lu et al. [23]	k-block diagonal regularizer	Clustering error	Real dataset	4000	The proposed method is for nonconvex when makes it challenging for optimization
6.	2017	P.A. Traganitis and G. B. Giannakis [24]	Sketched subspace clustering	Accuracy, time	Real datasets (HOPKIN 155)	2016	Few more evaluation measures should be taken. Statistical comparison is missing
7.	2017	Chen et al. [25]	Projection subspace clustering	Clustering accuracy	Real datasets	5920	Selecting efficient values of the parameter in the algorithm needs to be focused
8.	2016	D. Kumar et al. [26]	clusiVAT algorithm	Accuracy, time	Synthetic and real datasets	500	Comparison of algorithms is made on accuracy and run time only Lacks statistical test on algorithms
9.	2015	A. Kaur and A. Dutta [27]	Subscale algorithm	F1_Measure	Synthetic and real	6144	The only F1_measure is considered for performance evaluation Statistical comparison not made Comparison is made with few subspace clustering algorithms

10.	2014	Goebl et al. [28]	Optimal subspace clustering	Normalized mutual information (NMI)	Synthetic dataset	16	Only NMI used for evaluation of the proposed algorithm Maximum dimensions in high dimensional dataset are quite low, i.e., 16
11.	2014	A. Kaur and A. Dutta [29]	Subscale algorithm	F1_Measure	Synthetic and real	500	Other evaluation measures not considered Only 1 real dataset of 500 dimensions is considered
12.	2014	L. Lin et al. [30]	GA-PSO	Error rate	Real dataset	13	Low dimension dataset is considered for evaluation The proposed algorithm is not compared with standard subspace clustering algorithms
13.	2014	D. Kothari et al. [31]	FCM extended version using random sampling	Run time	Real dataset	54	Method for handling unloadable data is missing Not compared with standard subspace algorithm Low dimension data considered for evaluation
14.	2014	Fahad et al. [14]	Survey	Runtime, stability, internal validity	Real datasets	149	Few subspace clustering algorithms are considered. highest dimension evaluated in 149 only
15.	2014	Vidal & Favaro [32]	Low-rank subspace algorithm	Clustering error	Real dataset (motion segmentation)	2016	Evaluation of algorithms is made on clustering error only

(continued)

Table 1 (continued)

S. no.	Year	Authors and references	Approach used/proposed	Evaluation measure	Type of dataset	Maximum dimensions evaluated	Gaps
16.	2013	X. Chen et al. [33]	FG-k-means, a soft subspace clustering algorithm	F1_Measure, accuracy, precision, and recall	Synthetic and real datasets	500	Some real application is required for a testing algorithm Few more subspace algorithms should be considered for comparison
17.	2013	Cao et al. [34]	Weighting k-modes algorithm	Scalability, accuracy, and adjusted Rand index	Synthetic and real datasets	50	The proposed algorithm is evaluated on a maximum 50-dimensional data only Comparison is made with few algorithms only
18.	2013	Timmerman et al. [35]	Subspace K-means	Adjusted Rand index, cluster variance	Synthetic and real datasets	9	Lacks statistical comparison in algorithms Low dimension data is evaluated in the name of high dimensions
19.	2013	S. Jahirabadkar and P. Kulkarni [10]	Survey of subspace algorithms	None	None	None	Evaluation measure not defined Did not cover the complete list of subspace algorithms
20.	2013	Vijendra & Laxman [36]	Multi-objective subspace	Clustering error	Artificial and real datasets	250	Very few evaluation measures considered Performance assessment is made against two subspace algorithms only, i.e., Proclus and MOSCL

No.	Year	Author	Method	Metrics	Dataset	Dimensions	Remarks
21.	2012	Gajawada & Toshniwa [37]	Projected particle swarm optimization clustering algorithm	Number of mismatched dimensions in clusters	Synthetic dataset	126	Paper lacks in presenting the results in terms of standard evaluation measures. Not compared with any of the existing subspace clustering algorithms
22.	2012	Nourashrafeddin et al. [38]	EsubClus (evolutionary subspace clustering)	F1 score, running time	Synthetic dataset	500	Compared with the few existing subspace algorithms. Statistical comparison of algorithms is missing
23.	2011	Lu et al. [39]	Weighted PSO	F SCORE, average run time	Synthetic and real datasets	2000	Comparison of algorithms is made on F SCORE and run time only. Lacks statistical test on algorithms
24.	2011	Y. Zhao et al. [40]	Enhanced grid density-based approach for	Scalability, performance, and accuracy	Real datasets	100	Proposed algorithm could find subspaces for 100 dimension data only. The statistical test is missing
25.	2010	Chu et al. [7]	Density conscious subspace clustering	Execution time, precision, and recall	Synthetic and real datasets	16	Highest dimension considered for evaluation is 16 only. Statistical comparison is missing
26.	2009	Sun & Xiong [41]	Genetic algorithm -based high dimensional clustering	Error rate	Real life datasets	10	Very few dimensions are considered for performance evaluation. Evaluation metrics is error rate only
27.	2007	C. Bouveyron [6]	High dimensional data clustering model	BIC	Real and artificial datasets	100	Evaluated dimensions are up to 100 only

(continued)

Table 1 (continued)

S. no.	Year	Authors and references	Approach used/ proposed	Evaluation measure	Type of dataset	Maximum dimensions evaluated	Gaps
28.	2004	L. Parsons et al. [3]	Survey of subspace algorithms	Running vs. no. of instances/ dimensions	Synthetic dataset	100	Results of only two algorithms MAFIA and FINDIT were shown Evaluation measures were not discussed
29.	2004	L. Woo et al. [42]	FINDIT	F1_Measure, soundness	Synthetic datasets	50	The proposed algorithm was evaluated on different datasets but not compared with any of the clustering algorithms
30.	1998	P. S. Bradley, Fayyad, & C. Reina [43]	Scale K-means(SKM)	Log-likelihood, standard deviation	Synthetic and real datasets	100	Application of the proposed algorithm to the real-world problem is missing. Statistical analysis of the algorithm is lacking

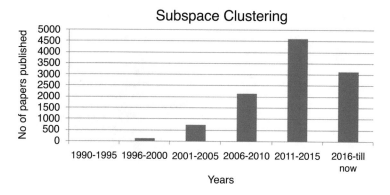

Fig. 7 Year-wise number of papers published on subspace clustering

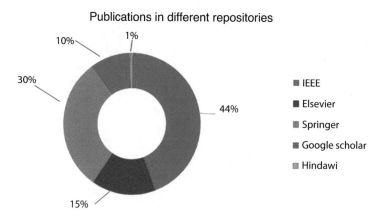

Fig. 8 Subspace clustering in different repositories

Real and synthetic datasets are obtained from [17]. Actual values of accuracy and F1_Measure of various subspace clustering algorithms on real datasets are extracted from [8]. While subscale algorithm is implemented in MATLAB R2013a platform adopting same parameter values given in [27]. Tables 2 and 3 shows the average rank and SRR rank of subspace clustering algorithms on real and synthetic datasets respectively. CLIQUE emerged to be on the first rank on accuracy, while MINECLUS best performs on F1_Measure.

Tables 4 and 5 present average and SRR ranks of subspace clustering algorithms on synthetic datasets in terms of F1_Measure and accuracy, respectively. Actual values of subspace algorithms on accuracy and F1_Measure on synthetic datasets were not available. Hence each algorithm is implemented on WEKA toolbox of subspace clustering provided by [17] with best parameter settings. It is observed that the subscale algorithm depicts better accuracy while DOC is best performer in terms of F1_Measure on synthetic datasets.

Fig. 9 Percentage of
papers published in
conferences and journals
on subspace clustering

Subspace Clustering Papers in Journals and Conferences
■ Conferences ■ Journals

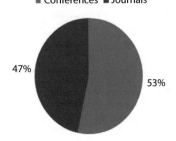

47%

53%

Table 2 Accuracy on real datasets

Accuracy		
	AR	SRR
SUBSCALE	9	11
CLIQUE	1	1
DOC	2	5
MINECLUS	4	7
SCHISM	3	2
SUBCLU	7	8
FIRES	11	10
INCY	4	3
PROCLUS	8	6
P3C	10	9
STATPC	6	4

Table 3 F1_Measure on real datasets

F1_Measure		
	AR	SRR
SUBSCALE	5	7
CLIQUE	9	8
DOC	2	3
MINECLUS	1	2
SCHISM	8	6
SUBCLU	7	9
FIRES	11	11
INCY	3	1
PROCLUS	6	4
P3C	10	10
STATPC	4	5

Table 4 Accuracy on synthetic datasets

Accuracy		
	AR	SRR
SUBSCALE	1	2
DOC	5	3
MINECLUS	6	4
SCHISM	3	7
FIRES	1	1
INCY	4	8
PROCLUS	8	6
STATPC	9	9
P3C	7	5

Table 5 F1_Measure on synthetic datasets

F1_Measure		
	AR	SRR
SUBSCALE	6	5
DOC	1	1
MINECLUS	4	2
SCHISM	7	7
FIRES	2	3
INCY	5	6
PROCLUS	3	4
STATPC	9	9
P3C	8	8

(ii) Scalability: Scalability of cell-based, density-based, and clustering-oriented-based subspace algorithms is shown in Figs. 9, 10, 11, 12, 13, and 14. Graphs are represented in terms of data dimensionality. Scalability depicts the performance of the algorithm with increasing number of dimensions. It is shown on synthetic datasets as number of records/objects are constant and dimensions vary from 10 to 75. Synthetic datasets are given in [17]. The X axis of graphs represents data dimensionality, and Y axis shows accuracy or F1_Measure.

Figures 10, 11, and 12 represent the performance of cell-based, density-based, and clustering-oriented-based algorithms, respectively, in terms of accuracy. It is observed from cell-based and density-based algorithms that SCHISM and INCY could not cope up after 25 dimensions and hence give results till D25 dataset only. MINECLUS accuracy varies randomly from dimensions to dimensions. DOC gives highest accuracy at 15 attribute dataset and then shows downfall. However, its accuracy shows slight improvement from D20 to D25 but again its performance declines. For density-based algorithms, FIRES shows improvement till D25, and then with

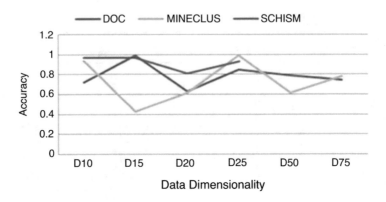

Fig. 10 Scalability of cell-based algorithms (accuracy)

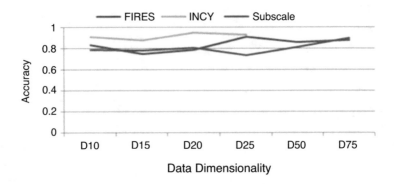

Fig. 11 Scalability of density-based algorithms (accuracy)

slight downfall, its performance becomes stagnant. SUBSCALE shows improvement in accuracy after D25 dimensional dataset. SUBSCALE and FIRES give highest and approximately same accuracy at 75-dimensional dataset. For clustering-oriented-based algorithms shown in Fig. 12, STATPC shows highest accuracy at D75 dataset. PROCLUS performance falls after 25-dimensional dataset. P3C shows low performance for overall datasets.

Figures 13, 14, and 15 depict the scalability of algorithms in terms of F1_ Measure. It has been noticed that subscale gives highest F1_Measure on the 75-dimensional dataset. However, the efficacy of DOC, FIRES, and P3C shows downfall in performance with an increase in dimensions.

It can be concluded from given experiments that on synthetic datasets, FIRES and DOC depict the best performance on the basis of accuracy and F1_Measure, while CLIQUE and MINECLUS are best performers on accuracy and F1_Measure on real datasets. That means cell-based and density-based algorithm is more appropriate to opt for subspace clustering.

The next section illustrates the various application areas along with future prospects.

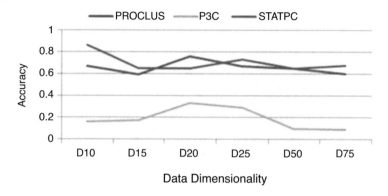

Fig. 12 Scalability of clustering-oriented-based algorithms (accuracy)

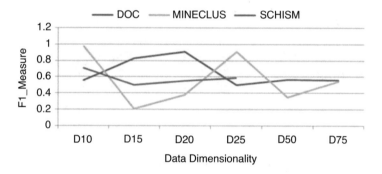

Fig. 13 Scalability of cell-based algorithms (F1_Measure)

7   Applications and Future Prospects

In previous sections, it has been shown that the trend of using subspace clustering algorithms for high dimensional problems is rising. This section discusses the application areas where subspace clustering algorithms are suitable. Additionally developing amalgamated subspace clustering algorithms are suggested. Following are the application areas of high dimensional clustering:

(i) Collaborative Filtering: The other name of collaborative filtering is a recommendation system. It is a social filtering technique where information is defined on basis of recommendations given by people [5]. People who like certain item in past are more likely to purchase in the future. People may like the items recommended by friends, neighbor, family, colleagues on social media, etc. Recommendations can be user-based or item-based. High dimensional clustering algorithms play an important role in such systems. The dataset is matrix of users and products. Clustering in such dataset retrieves the group of users liking the same product or group of items with relevant users. Hence subspace

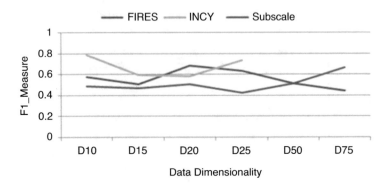

Fig. 14 Scalability of density-based algorithms (F1_Measure)

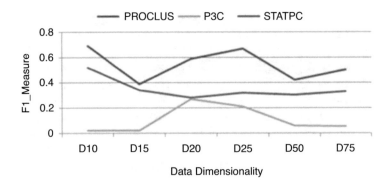

Fig. 15 Scalability of clustering-oriented-based algorithms (F1_Measure)

clustering could be applied [45, 46]. Some examples are Customer Recommendation System, Movie Recommendation System, etc.

(ii) Computer Vision: This field is based on mining useful information from single image or video to attain automatic visual understanding. The attributes extracted from image or video are large in number. An example of such dataset is image segmentation data [47]. The dataset contains 2310 data objects and 19 attributes. Clustering techniques are applied to cluster in either shape group or RGB group [33]. Similarly there a number of datasets with large attributes where clusters may exist in different subspaces [48]. Field of computer vision where subspace clustering can be applied is Facial Recognition, Gesture Recognition, etc.

(iii) Biological Dataset: Gene expression dataset is the most widely used dataset where subspace clustering can be applied [5]. Microarray DNA is a technology which measures a large amount of genes expressions under different circumstances. In order to understand the various types of diseases caused by genetic disorder at different levels, subspace clustering is required [49].

(iv) Text Documents: Clustering in text documents is an important task for web mining. Web pages are clustered on the basis of frequency of terms occurring in the page. Text documents are represented as high dimensional feature vector, where each feature is a frequency of term in document. Each document is represented by a data record/object. Hence the dataset formed from text document is high dimensional data [50]. The cluster of related document may exist on basis of similar word count, themes, etc. Hence subspace clustering techniques are applicable in such datasets.

(v) Distributed Databases: Massive amount of data is being dissipated by a number of sources like social media, microarray DNA, etc. Nowadays such a large amount of data is stored in different physical locations named distributed databases. The massive data is fragmented so that it can be distributed over multiple servers and parallel queries can be executed. Fragmentation can be row-wise (horizontal) or column-wise (vertical). Subspace clustering techniques can be applied to aid the fragmentation process. However, no work has been done yet on this application using subspace clustering approach.

(vi) Social Network: The data produced from social media is large in volume. Clustering can be performed on the social network to find influential groups in the network. This task is achieved by analyzing the linkages (edges) and nodes of the network. Clustering can be done on bases of various attributes associated with each node depending upon the objective in hand. Determining influential groups on the basis of various topics can be helpful in promotional activities, market segmentation problem, community detection problem [51], etc. Hence subspace clustering can be applied to find the groups in subsets of dimensions. However, a little amount of work is done on this application using subspace clustering approach.

In recent years, subspace clustering is being used with metaheuristic approaches. Nature-inspired algorithms [52, 53] are the prominent metaheuristic techniques that are used for determining the near-optimal solution to complex and hard problems. The first hybrid approach of subspace clustering with an evolutionary algorithm is proposed in [54]. With the advent of new metaheuristic algorithms, subspace clustering results could be improved [30, 36]. Some other algorithms can be developed amalgamating artificial bee colony algorithm, grey wolf algorithm, etc., with various subspace clustering algorithms. The hybrid algorithms developed were not applied to the applications described above. This will give a new direction of future work to researchers in the field of subspace clustering of high dimensional data.

8 Conclusion

Subspace clustering finds the clusters existing in various subsets of dimensions. The chapter presents a comprehensive survey on subspace clustering approaches, evaluation metrics, and application areas. The chapter also reveals the significance of subspace clustering in literature by presenting the statistical data. Comparison of conventional subspace clustering algorithms is also depicted through average ranking and success rate ratio ranking. Performance assessment of algorithms is made through scalability on basis data dimensionality. The chapter answers the following research questions:

(i) What are the major challenges faced by traditional clustering algorithms to cluster high dimensional data?
    ANSWER: The problem of the curse of dimensionality is described in the first paragraph of Sect. 1.
(ii) What search techniques are being used in subspace clustering to determine subspaces?
    ANSWER: Top-down and bottom-up search techniques used to find subspaces and are described in Sect. 3.
(iii) What are evaluation measures for comparing subspace clustering algorithms?
    ANSWER: Different evaluation measures are described in Sect. 4.
(iv) What is the current scenario of subspace clustering?
    ANSWER: Literature survey on subspace clustering with statistical data is illustrated in Sect. 5.
(v) What are the research gaps in the literature and the future prospects of subspace clustering?
    ANSWER: Research gaps are given in Table 1, and future prospects are presented in Sect. 7.

Thus, the chapter is useful to the researchers planning to work in the field of subspace clustering. Additionally, it suggests the algorithms to develop in the future along with application areas.

References

1. J. Han, J. Pei, M. Kamber, *Data Mining: Concepts and Techniques* (Elsevier Inc, 2011)
2. I. Assent, Clustering high dimensional data. Wiley Interdiscip. Rev. Data Min. Knowl. Discov. **2**, 340–350 (2012). https://doi.org/10.1002/widm.1062
3. L. Parsons, E. Haque, H. Liu, Subspace clustering for high dimensional data: A review. ACM SIGKDD Explor. Newsl. **6**, 90–105 (2004). https://doi.org/10.1145/1007730.1007731
4. H.-P. Kriegel, P. Kroger, M. Renz, S. Wurst, A generic framework for efficient subspace clustering of high-dimensional data. ICDM, 250–257 (2005)
5. H.-P. Kriegel, P. Kröger, A. Zimek, Clustering high-dimensional data: A survey on subspace clustering, pattern-based clustering, and correlation clustering. ACM Trans. Knowl. Discov. Data **3**, 1–58 (2009). https://doi.org/10.1145/1497577.1497578
6. C. Bouveyron, S. Girard, C. Schmid, High-dimensional data clustering. Comput. Stat. Data Anal. **52**, 502–519 (2007). https://doi.org/10.1016/j.csda.2007.02.009
7. Y.H. Chu, J.W. Huang, K.T. Chuang, D.N. Yang, M.S. Chen, Density conscious subspace clustering for high-dimensional data. IEEE Trans. Knowl. Data Eng. **22**, 16–30 (2010). https://doi.org/10.1109/TKDE.2008.224

8. E. Müller, S. Günnemann, I. Assent, T. Seidl, Evaluating clustering in subspace projections of high dimensional data, in *Proceedings of the VLDB Endowment*, (2009), pp. 1270–1281. https://doi.org/10.14778/1687627.1687770
9. A. Patrikainen, M. Meila, Comparing subspace clusterings. IEEE Trans. Knowl. Data Eng. **18**, 902–916 (2006). https://doi.org/10.1109/TKDE.2006.106
10. S. Jahirabadkar, P. Kulkarni, Scaf – An effective approach to classify subspace clustering algorithms. Int. J. Data Min. Knowl. Manag. Process. **3**, 69–86 (2013)
11. P. Brazdil, C. Soares, A comparison of ranking methods for classification algorithm selection. Mach. Learn. ECML 2000 **1810**, 63–75 (2000). https://doi.org/10.1007/3-540-45164-1_8
12. C. Bouveyron, C. Brunet, Model-based clustering of high-dimensional data: A review. Comput. Stat. Data Anal. Elsevier, 52–78 (2013)
13. M. Steinbach, E. Levent, V. Kumar, The challenges of clustering high dimensional data. New Vistas Stat. Phys. Appl. Econophys. Bioinf. Pattern Recognit., 273–309 (2004). https://doi.org/10.1007/978-3-662-08968-2_16
14. A. Fahad, N. Alshatri, Z. Tari, A. Alamri, I. Khalil, A.Y. Zomaya, S. Foufou, A. Bouras, A survey of clustering algorithms for big data: Taxonomy and empirical analysis. IEEE Trans. Emerg. Top. Comput. **2**, 267–279 (2014). https://doi.org/10.1109/TETC.2014.2330519
15. P. Agarwal, S. Mehta, Comparative analysis of nature inspired algorithms on data clustering. IEEE Int. Conf. Res. Comput. Intell. Commun. Netw., 119–124 (2015)
16. P. Agarwal, S. Mehta, Enhanced flower pollination algorithm on data clustering. Int. J. Comput. Appl. Taylor Fr. **7074**, 144–155 (2016). https://doi.org/10.1080/1206212X.2016.1224401
17. E. Müller, S. Günnemann, I. Assent, T. Seidl, I. Färber, Evaluating clustering in subspace projections of high dimensional data, http://dme.rwth-aachen.de/en/OpenSubspace/evaluation
18. I.A. Sarafis, P.W. Trinder, A.M.S. Zalzala, Towards effective subspace clustering with an evolutionary algorithm. 2003 Congr. Evol. Comput. CEC 2003 – Proc. **2**, 797–806 (2003). https://doi.org/10.1109/CEC.2003.1299749
19. P. Ji, I. Reid, R. Garg, H. Li, M. Salzmann, Adaptive low-rank kernel subspace clustering. arXiv:1707.04974 [cs.CV] (2017)
20. J. Yang, J. Liang, K. Wang, P. Rosin, M.-H. Yang, Subspace clustering via good neighbors. IEEE Trans. Pattern Anal. Mach. Intell., 1 (2019). https://doi.org/10.1109/tpami.2019.2913863
21. D.V. Paul, C.S. Nayagam, Use of subspace clustering algorithm for students' competency and subject knowledge assessment. Int. J. Knowl. Syst. Sci. **9**, 70–83 (2018). https://doi.org/10.4018/IJKSS.2018040104
22. M.C. Tsakiris, R. Vidal, Theoretical analysis of sparse subspace clustering with missing entries. 35th Int. Conf. Mach. Learn. ICML 2018 **11**, 7940–7949 (2018)
23. C. Lu, J. Feng, Z. Lin, T. Mei, S. Yan, Subspace clustering by block diagonal representation. IEEE Trans. Pattern Anal. Mach. Intell. **41**, 487–501 (2019). https://doi.org/10.1109/TPAMI.2018.2794348
24. P.A. Traganitis, G.B. Giannakis, Sketched subspace clustering. IEEE Trans. Signal Process. **2018**, 1–18 (2017). https://doi.org/10.1109/TSP.2017.2781649
25. X. Chen, M. Liao, X. Ye, Projection subspace clustering. J. Algorithms Comput. Technol. **11**, 224–233 (2017). https://doi.org/10.1177/1748301817707321
26. D. Kumar, J.C. Bezdek, M. Palaniswami, S. Rajasegarar, C. Leckie, T.C. Havens, A hybrid approach to clustering in big data. IEEE Trans. Cybern. **46**, 2372–2385 (2016). https://doi.org/10.1109/TCYB.2015.2477416
27. A. Kaur, A. Datta, A novel algorithm for fast and scalable subspace clustering of high-dimensional data. J. Big Data. **2**, 17 (2015). https://doi.org/10.1186/s40537-015-0027-y
28. S. Goebl, H. Xiao, C. Plant, C. Bohm, Finding the optimal subspace for clustering, in *IEEE International Conference on Data Mining (ICDM)*, (2014), pp. 130–139. https://doi.org/10.1109/ICDM.2014.34
29. A. Kaur, A. Datta, SUBSCALE: Fast and scalable subspace clustering for high dimensional data, in *IEEE International Conference on Data Mining Workshops (ICDM)*, (2014), pp. 621–628. https://doi.org/10.1109/ICDMW.2014.100

30. L. Lin, M. Gen, Y. Liang, A hybrid EA for high-dimensional subspace clustering problem. Proc. 2014 IEEE Congr. Evol. Comput. CEC 2014, 2855–2860 (2014). https://doi.org/10.1109/CEC.2014.6900313
31. D. Kothari, S.T. Narayanan, K.K. Devi, Extended fuzzy C-means with random sampling techniques for clustering large data. Int. J. Innov. Res. Adv. Eng. **1**, 1–4 (2014)
32. R. Vidal, P. Favaro, Low rank subspace clustering (LRSC). Pattern Recogn. Lett. **43**, 47–61 (2014). https://doi.org/10.1016/j.patrec.2013.08.006
33. X. Chen, Y. Ye, X. Xu, J.Z. Huang, A feature group weighting method for subspace clustering of high-dimensional data. Pattern Recogn. **45**, 434–446 (2012). https://doi.org/10.1016/j.patcog.2011.06.004
34. F. Cao, J. Liang, D. Li, X. Zhao, A weighting k-modes algorithm for subspace clustering of categorical data. Neurocomputing **108**, 23–30 (2013). https://doi.org/10.1016/j.neucom.2012.11.009
35. M.E. Timmerman, E. Ceulemans, K. De Roover, K. Van Leeuwen, Subspace K-means clustering. Behav. Res. Methods **45**, 1011–1023 (2013). https://doi.org/10.3758/s13428-013-0329-y
36. S. Vijendra, S. Laxman, Subspace clustering of high-dimensional data: An evolutionary approach. Appl. Comput. Intell. Soft Comput. **2013**, 1–13 (2013)
37. S. Gajawada, D. Toshniwal, Projected clustering using particle swarm optimization. Procedia Technol. **4**, 360–364 (2012). https://doi.org/10.1016/j.protcy.2012.05.055
38. S. Nourashrafeddin, D. Arnold, E. Milios, An evolutionary subspace clustering algorithm for high-dimensional data. Proc. Fourteenth Int. Conf. Genet. Evol. Comput. Conf. Companion, 1497–1498 (2012). https://doi.org/10.1145/2330784.2331011
39. Y. Lu, S. Wang, S. Li, C. Zhou, Particle swarm optimizer for variable weighting in clustering high-dimensional data. Mach. Learn. **82**, 43–70 (2011). https://doi.org/10.1007/s10994-009-5154-2
40. Y. Zhao, J. Cao, C. Zhang, S. Zhang, Enhancing grid-density based clustering for high dimensional data. J. Syst. Softw. **84**, 1524–1539 (2011). https://doi.org/10.1016/j.jss.2011.02.047
41. H. Sun, L. Xiong, Genetic algorithm-based high-dimensional data clustering technique, in *International Conference on Fuzzy Systems and Knowledge Discovery*, (2009), pp. 485–489. https://doi.org/10.1109/FSKD.2009.215
42. K.G. Woo, J.H. Lee, M.H. Kim, Y.J. Lee, FINDIT: A fast and intelligent subspace clustering algorithm using dimension voting. Inf. Softw. Technol. **46**, 255–271 (2004). https://doi.org/10.1016/j.infsof.2003.07.003
43. P.S. Bradley, U. Fayyad, C. Reina, Scaling clustering algorithms to large databases. KDD-98, 1–7 (1998)
44. C.F. Olson, D.C. Hunn, H.J. Lyons, Efficient Monte Carlo clustering in subspaces. Knowl. Inf. Syst. **52**, 751–772 (2017). https://doi.org/10.1007/s10115-017-1031-7
45. L. Boratto, S. Carta, Using collaborative filtering to overcome the curse of dimensionality when clustering users in a group recommender system, in *Proceedings of the 16th International Conference on Enterprise Information Systems*, (2014), pp. 564–572. https://doi.org/10.5220/0004865005640572
46. U. Kuzelewska, Clustering algorithms in hybrid recommender system on MovieLens data. Stud. Logic. Gramm. Rhetor. **37**, 125–139 (2014). https://doi.org/10.2478/slgr-2014-0021
47. K. Bache, M. Lichman, *UCI Machine Learning Repository* (University of California, School of Information and Computer Science, Irvine, 2006)
48. R. Vidal, Subspace Clustering. IEEE Signal Process. Mag. **28**, 52–68 (2011). https://doi.org/10.1109/MSP.2010.939739
49. D. Jiang, C. Tang, A. Zhang, Cluster analysis for gene expression data: A survey. IEEE Trans. Knowl. Data Eng. **16**, 1370–1386 (2004)
50. L. Jing, M.K. Ng, J. Xu, J.Z. Huang, Subspace clustering of text documents with feature weighting K-means algorithm. Lect. Notes Comput. Sci, 802–812 (2005). https://doi.org/10.1007/11430919_94

51. Z. Zhao, S. Feng, Q. Wang, J.Z. Huang, G.J. Williams, J. Fan, Topic oriented community detection through social objects and link analysis in social networks. Knowledge-Based Syst. **26**, 164–173 (2012). https://doi.org/10.1016/j.knosys.2011.07.017

52. P. Agarwal, S. Mehta, Nature-inspired algorithms: State-of-art, problems and prospects. Int. J. Comput. Appl. **100**, 14–21 (2014). https://doi.org/10.5120/17593-8331

53. P. Agarwal, S. Mehta, Empirical analysis of five nature-inspired algorithms on real parameter optimization problems. Artif. Intell. Rev., 1–57 (2017). https://doi.org/10.1007/s10462-017-9547-5

54. H. Road, S. Jose, Automatic subspace clustering mining of high dimensional applications for data. Proc. 1998 ACM SIGMOD Int. Conf. Manag. Data **27**, 94–105 (1998). https://doi.org/10.1145/276305.276314

Ant Colony Optimization Technique in Soft Computational Data Research for NP-Hard Problems

Moha Gupta, Puneet Garg, and Prerna Agarwal

1 Introduction

On recollecting past, in 1990, a new trend was started in the area of designing the consumer electronics, household appliances and various other consumer products used widely. A new term machine intelligence quotient (MIQ) also came into being which marks the increase in all these consumer products as compared to before 1990. And as the technology rises and more intelligent systems came into being, products such as microwaves and washing machines came which focuses on some particular tasks assigned to them; they prove that machines can make intelligent decisions and also learn from their experience. There are numerous factors which causes the increase in MIQ. Zadeh (1996) studied that most important factor affecting in soft computing or can be called as fuzzy logic which enables the machine to mimic as humans approximately. Some factors which caused the increase in high MIQ are based on the observations such as F.L. Smidth cement kiln, and the Sendai system was designed by the very renowned company Hitachi which made a groundwork for the fuzzy logic for designing and production of increase in MIQ products [1]. Another product which a shower head controlled by fuzzy logic was made in 1987 by Matsushita; they also made fuzzy controlled washing machine in 1989.

M. Gupta · P. Garg
J. C. Bose University of Science and Technology, YMCA, Faridabad, Haryana, India

P. Agarwal (✉)
JEMTEC, Greater Noida, India

2   Soft Computing

Soft computing is the fusion of three main computing models, viz., fuzzy logic, the neural networks, and evolutionary computation including the probabilistic reasoning and genetic algorithm for providing the application that can be used in the real world [2]. Soft computing is called as a synergistic integration which is referred as computational intelligence [3, 4]. The mentioned computing models, i.e., fuzzy logic, neural network, and probabilistic reasoning, share some similar features and functions and are complementary to each other. These models can be combined and used after that for more effective results and functioning. Hybrid computing system is made by combining hard and soft computing [5]. In traditional system, hard computing was used which is the best technique to solve the mathematical problems which are not related to real world on the other hand soft computing is good in solving real world issues. Hard computing uses binary logic and some predefined functions, whereas soft computing uses fuzzy system and probabilistic reasoning used for the human brain. The most popular and successful combination which was first made was the neurofuzzy hybrid system [6] and geneticfuzzy system [7].

A. Genetic Algorithms (GA)

They are an algorithm used for optimization which involves iterative search procedures that are based on Darwinism (process of natural selection) and the evolutionary genetics [8]. John Holland (Professor) from University of Michigan proposed these algorithms [9], and later more contributions were added by Goldberg [10]. GA's main objective is to optimize the input variables known as "fitness function." To perform the same, it maintains a population which is called as "individuals." Five phases are there in a genetic algorithm:

1. Initial population – the phase starts from a set of individuals known as population. That individual is differentiated from others using some parameters or variables called as "Genes." Then these genes are formed "Chromosome" by joining the genes and are represented using string (usually binary values).
2. Fitness function – it defines about how fit is an individual which is calculated by assigning a fitness score to every individual.
3. Selection – in this phase, the best fit individual is selected and their genes are passed to next generation.
4. Crossover – in this phase, the individuals or parents selected with highest fitness score are mated, and a crossover point is selected randomly and the offspring are created.
5. Mutation – in mutation, some of the offspring having low random probability, their bits are flipped in order to maintain diversity in the population and to avoid premature convergence.

B. Neural Network

Neural network represents a series or pattern of algorithm which helps in building relationship among the huge amount of data in a way human brain does. On an average, human brain has 3 * 1010 neurons of different types and each neuron making a connection to around 104 synapses [11]. Neural network consists of numerous neurons, where each neuron is working differently and simultaneously. The neurons have variable weights and are connected via synapses [12]. The nodes of the neurons take input variable and gives result by performing simple operations on the input variables. Then these results are moved further to next neuron, the output is known as activation value of each node. In the 1940s, some researchers named McCulloch and Pitts proved that neurons can be molded in a way that they act as a device with simple threshold which can perform logic functions [13]. Also in 1940 after them, Hebb gave the Hebbian rule which describes the synaptic learning between neurons [14]. Many researchers contributed in this filed proposing variations in neural networks and its types and more details. Rosenblatt gave the perceptron model [15], while Widrow and Hoff gave Adaline model [16] which was trained using Least Mean Square method (LMS). The popularity in neural network decreased after Minsky and Papert proved mathematically that perceptron doesn't perform good in complex logic functions [17].

C. Fuzzy Logic

Fuzzy logic and fuzzy set theory originated in 1994 by Lofti Zadeh which captures uncertainty or ambiguity [18–21]. Broadly, fuzzy logic and fuzzy set theory are synonym of each other [1]. Fuzzy set theory is a theory of unsharp boundary classes, while fuzzy logic is a branch of fuzzy set theory. Some other branches include fuzzy mathematical programming, fuzzy arithmetic, fuzzy graph theory, fuzzy data analysis, and fuzzy topology. Two concepts in fuzzy logic play an important role:

1. Linguistic variable – It is a variable who have values in the form of words or sentences in a synthetic or natural language.
2. Fuzzy if-then rule – Both if and then are propositions which consists of linguistic variables. Apart from Zadeh, Joseph Goguen also contributed in the field of fuzzy logic and received great success with the development of "logic of inexact concepts" [22]. After this theory and achievement, fuzzy logic started to be applied in various fields.

3 Applications of Soft Computing

From over the years, soft computing had gained an increase, and it not only focused on theoretical areas but also in the field of real-world problems and are able to solve them efficiently. A number of journals and books have been already published in the area of soft computing and fuzzy logic [23–35]. Some of the application areas are discussed in Fig. 1.

Fig. 1 Applications of soft computing

A. Mathematics (Pure and Applied)

Some areas in mathematics are redesigned after the success of soft computing and fuzzy logic such as resituated lattices, differential equations, approximation spaces, topology, tolerance spaces, algebraic study dealing in imprecision, vagueness, and uncertainty [36].

B. Biomedical

Biomedical includes both medicine and biology. Soft computing helped in advancement of healthcare by making a bridge between medical and engineering. It improved the treatment in areas of therapy, diagnosis, and monitoring the diseases.

C. Computer Science /IT

It also includes many fields of electrical, software, and hardware systems. Soft computing plays an important role in designing of microprocessors, supercomputers, circuits, and software and hardware systems. On incorporating every part into one single bigger unit, soft computing focuses on the working of computer system.

D. Case-Based Reasoning

It includes understanding, learning, and solving of a problem using the memory processes. Soft computing uses the old or past information and learning to justify the new demands and solutions. Nowadays, soft computing is working on real-world problems of case-based reasoning [37].

E. Multimedia

Multimedia processing requires high learning and strong cognitive abilities, and for this, soft computing is the best possible way which is getting very popular. Soft computing has a wide range of applications in multimedia such as color quantization, video editing and sequencing, image processing, image analysis and segmentation, biometrics, and document processing. The area of image processing is used in scientific, commercial, and many other research areas [38].

F. Environmental Science

The main objective of Environment Engineering is saving the natural environment such as water, land, and air resources to get clean and sanitized resources for humans and for flora and fauna; it will also decrease pollution.

G. Data Mining

It helps in finding patterns among the large data used in artificial intelligence, statistical models, and machine learning areas. Soft computing helps in extracting relevant and useful information from the large dataset and make it understandable so that it can be used for real-world problems.

H. Mechanical Engineering

The Concepts of Mechanical Engineering may be applied to both physics and material sciences. The concepts are used for analysis, designing, manufacturing, and maintaining the systems. Soft computing helps in determining proper use of heat and the mechanical powers used for designing and operating the tools and machines.

4 Ant Colony Optimization Technique (ACO)

Optimization is a process of maximizing or minimizing the function or the process in order to get optimized cost or output given some inputs. The optimization algorithm has some properties as scalability, generalization, and efficiency in its execution process for large datasets. The optimization techniques are broadly classified into continuous optimization, combinatorial optimization, alternating minimization, and branch and bound. They are further divided into different types of optimization techniques such as linear, quadratic, heuristic, metaheuristic, global method, neighborhood-based algorithm, simulated annealing, ant colony optimization, particle swarm optimization, genetic algorithm, and many more. In this chapter, the variations and developments of ant colony optimization technique is discussed which is used to solve a NP-hard problem, i.e., travelling salesman, job-shop scheduling problem, and quadratic assignment problem.

There are numerous ways and techniques possible to solve the NP-hard problems such as particle swarm optimization [42, 43] and genetic algorithm, but the mainly used technique is metaheuristic, i.e., ant colony optimization technique [40, 41]. These techniques are used to get the optimal values. The meta-heuristic is one of the most used and most effective method of using ACO to solve TSP. Stutzle and Dorigo explored a way to solve TSP using meta-heuristic ACO [44–46]. The ACO algorithm represents the ant behavior for getting the smallest path in the real world. It was proposed in 1991 [47]. The ant starts from one side of the node and moves toward another node forming a route; this process is repeated while selecting the next node using probability function. The node with minimum distance and maximum number of pheromones is selected and that node will be visited [48].

A tabulist is prepared which consists the record of all nodes that are visited by ants to avoid the repeated visit to those nodes as all the nodes need to be covered. The tabulist is full when all the nodes are covered by ants. After the tabulist is full, pheromone update rules are used, and the loss of pheromone is calculated. The loss of pheromone will be more for shorter routes as more pheromones are left unvisited, while for longer routes, loss will be less. This process is repeated till the system goes into a stable or stagnant situation, i.e., no need to look for alternative routes [49]. The ants randomly start searching for the food and while they are going back to their colony the pheromone starts depositing. If some ant finds the optimal or shorter path, then other ants also start following the that shorter path as shown in Fig. 2.

4.1 Mathematical Model of ACO

The ant colony optimization algorithm was developed mainly for the TSP problem. The focus and main part of this algorithm are ants. They are the agents which construct solutions iteratively and individually. The solutions are performed step-by-step staring from zero, i.e., empty solution. The ants move from one state to another trying to complete the solution. The process is completed when the ant completes the solution. At each step, the ant calculates the number of feasible and possible states from that state, and probabilistically the next state is selected from those states by the ant [59]. These feasible solutions are selected according to pheromone intensity. The probability includes two functions: d_{ij} (path length) and τ_{ij}

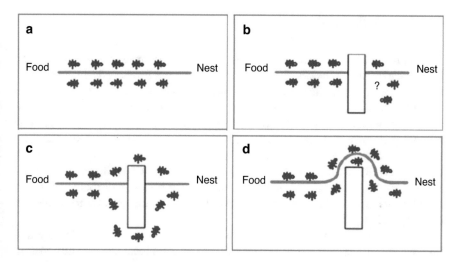

Fig. 2 (**a**) Ants in pheromone trail, (**b**) obstacle between trails, (**c**) ants discovers two paths and crosses the obstacle, (**d**) a new shorter path discovered forming pheromone trail

(pheromone on the path) [60]. While updating the pheromone after completion of the cycle by ant, the path of the pheromone is changes. The probability of state transition $p_{ij}^{k}(t)$ and updation of pheromone equation Tij are:

$$p_{ij}^{k}(t) = \frac{\left(Tij(t)\right)\alpha\left(\eta ij(t)\right)\beta}{\sum\limits_{s \in J_k^{j}(i)} \left(Tis(t)\right)\alpha\left(\eta is(t)\right)\beta}$$
(1)

$$\textbf{\textit{If } } j \in J_k^{(i)}, \text{ or else } 0.$$

The above equation is used to calculate state transition probability, where, α, β are the degree of the pheromone, $p_{ij}^{k}(t)$ is the probability of ant "k" from state i to j, and η_{ij} is heuristic factor. Pheromone update equations are calculated as:

$$T_{ij}(t+n) = \rho * Tij(t) + \Delta Tij$$
(2)

$$\Delta Tij = \sum_{k=1}^{m} \Delta Tij^{k}$$
(3)

$$\Delta Tij^{k} = \frac{Q}{L_k}, \text{ if an } tk \text{ travels through } ij \text{ state else } 0$$
(4)

In the above equations, ΔTij is current iteration of pheromone, Lk is path length and ρ is coefficient of the pheromone persistent.

5 Variations in ACO

ACO is being used to solve TSP. The first ACO algorithm was Ant System (AS) [47], made by three scholars in 1991. The algorithm has its own advantage and disadvantage as the algorithm is robust but due its slow convergence and problem in expanding of search space, it is not used now [41]. Many new algorithms are developed after this such as Elitist AS [50], the max-min AS [51], ACO [48], novel max-min AS [52], rank-based AS [53], adaptive dynamic AS [54], improved ACO [56], moderate AS [55], cooperative genetic AS [57], hybrid system of ACO and GA (ACO-GA) [58], hybrid of ACO and cuckoo search (ACO-CS) [58], and many more hybrid systems. The elitist AS focuses mainly on the best path or route possible to improve global searching of ant system. But still it is not able to solve the drawback of ant system, i.e., slow convergence.

After elitist AS, ant colony system was made to increase the efficiency and improve performance of the system by using some extra local pheromone updating with already used updation in pheromone at the end of construction phase [48]. To solve the problem of slow convergence, rank-based AS was developed which uses ranking strategy by weighing the ant's contribution up to trail level [53]. This algorithm was only able to improve the convergence problem a little; hence the max-min AS was developed after this. The max-min AS tried to avoid the premature convergence [51] by improving the pheromone updation method. It allows only the best ant to update the trails of pheromone. But, this system also not solves the problem, so the novel max-min AS [52] was used to improve some parameters used but was unable to solve the problem. While the hybrid systems are though complex to perform and calculate yet they proved to get better results than the other algorithms used.

6 Some Common Computational Problems Solved Using ACO

ACO helps in solving some of the complex problems which are categorized into following broad categories [61] such as scheduling problems which include jobshop scheduling problem (JSP), group-shop scheduling problem (GSP), open-shop scheduling problem (OSP), and resource-constrained project scheduling problem (RCPSP). Other category is vehicle routing problem which includes travelling salesman problem (TSP), period vehicle routing problem (PVRP), multi-depot vehicle routing problem (MDVRP), split delivery vehicle routing problem (SDVRP), vehicle routing problem with time windows (VRPTW), and stochastic vehicle routing problem (SVRP). Another category is assignment problem which includes generalized assignment problem (GAP), quadratic assignment problem (QAP), redundancy allocation problem, and frequency assignment problem (FAP).

Another category is set problem which has following subproblems including set partition problem (SPP), set covering problem (SCP), maximum independent set problem (MIS), and multiple knapsack problem (MKP). Some other categories are data mining, protein folding, image processing, system identification, and many more. Some most common problems solved by ACO are discussed in the following sections.

6.1 Travelling Salesman Problem

TSP is a NP-hard optimization problem. The main aim of TSP is to find a graph which is closed and keeping in mind that it covers minimum length. The salesman starts from one location and comes back again to the starting location via covering all the locations in between, but the condition is that one location to be visited only once as shown in Fig. 3. It can be calculated as [39]:

n: number of locations to be visited
d_{ab}: distance from location a to location b
v_a: extra variables used while visiting location a

Decision variable is defined by:

$$X_{ab} = \begin{cases} 1, \text{if travelled from location } a \text{ to } b \\ 0, \text{else} \end{cases} \qquad (5)$$

The main aim of TSP is finding the minimum total distance travelled from one location to another, which is given by:

$$min.Z = \sum_{a}^{n} \binom{n}{a=1} \sum_{b=1}^{n} d_{ab} X_{ab} \qquad (6)$$

When using ACO to solve TSP, $Т_{ij}$ (t) (pheromone strength) is associated to every arc, and this pheromone strength is updated after every t iteration counter. When ACO is used to solve symmetric TSP, then $Т_{ij}$ (t) = $Т_{ji}$ (t), but when applied

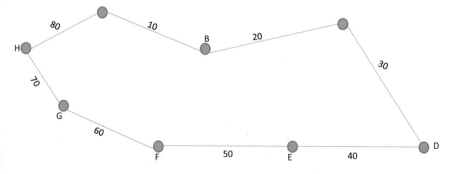

Fig. 3 TSP with 8 cities (A-B node nos.) and each having distance (d) between them

to asymmetric TSP, then they are not equal, i.e., $\mathrm{T}ij\,(t) \neq \mathrm{T}ji\,(t)$ [62]. Initially, "*m*" ants are placed on randomly chosen cities and then state transition rules are applied. The ant constructs the tour according to the pheromone values. The ant selects a city from *i* to *j* probabilistically, according to the pheromone strength values. They usually prefer the cities closer and more connected to the previous city and also having pheromone strength. Each ant constructs a tabulist which is a kind of memory, it is used for determining which cities are visited and which cities need to be visited. This tabulist helps in getting a feasible solution. As the ant proceeds to cover the route and pheromones are updated, the city which is visited maximum number of times by the ant is given higher probability. In this way, it becomes easy for the ant to select next city with high probability pheromone strength. Mainly, the algorithm used for solving NP-hard problems follows the following algorithm [48, 63–66]:

Generally, all algorithms used to solve TSP follow some general steps as shown in Fig. 4. The procedure of updation of pheromone and visiting the next city and this process continues until the termination condition is reached.

6.2 Job-Shop Scheduling Problem (JSP)

JSP is the most difficult problem considered in scheduling theory which is NP-hard [67]. The main aim of JSP is to get a schedule which has the minimum length. It is composed of set *P* of *j* Jobs and set *Q* of m machines. Each and every job need to be executed on each and every machine, and it contains *m* operations oi1, ..., oim. There are total *j* * *m* operations which are run taking processing time as pik. In JSP, each machine can take only one job and each job can be run on only one machine simultaneously. As the objective of JSP is to minimize length, Lmax parameter is used which is used to calculate performance measure that needs to be minimized. The starting time (tik) is determined for every operation to get minimal value and is calculated as [68]:

$$\text{Min}\{L_{max}\} = \min\left(\text{feasible solution}\right)\left\{\max\{t_{ik} + p_{ik}\} : \forall\, P_i\, TP, \forall\, Q_k\, TQ\right\} \quad (7)$$

These are subject to some constraints as [69]:

Fig. 4 TSP pseudocode using ACO

procedure

set the parameters and initialize the pheromone trails

while (termination condition is not met) do

Construct solutions

Apply the local search method *(optional)*

Update the trails

End

Start time $t_{ik} >= 0$
Precedence constraint $t_{ik} - t_{ih} >= p_{ik}$, when o_{ih} preceding o_{ik}
Disjunctive restriction $t_{pk} - t_{ik} + K(1 - y_{ipk}) >= p_{ik}$, $y_{ipk} = 1$, if o_{ik} preceding o_{pk}
$\qquad t_{ik} - t_{pk} + K(y_{ipk}) >= p_{ik}$, $y_{ipk} = 0$, in other cases

the constraints of JSP are discussed:

Start time – as the time of starting is not fixed and specified, it can start at any time whenever machine is available.

Precedence constraint – this constraint as needed as the next job cannot start processing until the first job has completed its work which is predefined, so as to prevent two jobs working simultaneously.

Disjunctive restriction – one machine can work for only one job at a time, so this constraint is applied so that one job to be completed on one machine and till then no new job to be given to that machine even when the jobs are waiting.

6.3 Quadratic Assignment Problem

The QAP was first developed by Beckmann and Koopmans (1957), belonging to the combinatorial optimization problems. It is one of the mathematical model used for economical activities. The main aim of QAP is to minimize cost of assignments having n locations and n facilities. The n facilities are being assigned to the n locations. The QAP is represented using three matrices of n x n dimensions as follows:

$D = [d_{ih}]$ = matrix of distance between the locations i and h.
$F = [f_{jk}]$ = matrix of flow between activities j and k.
$C = [c_{ij}]$ = matrix of cost of assignment between locations j to i.

Matrix D and matrix F are symmetrical matrices which are integer valued and the matrix values of c_{ij}, i.e., assignment cost is ignored because they make only minor significance which can be ignored. The cost between activities for transferring of the data is expressed by the product of distance between locations, $d_{ih} f_{\pi(i)\pi(h)}$. To fulfil the objective of QAP, i.e., minimizing cost the following equation is used:

$$\min z = \sum_{i,h=1}^{n} d_{ih} f_{\pi(i)\pi(h)} \tag{8}$$

A permutation matrix is obtained of $n \times n$ dimension, matrix Xij, the value is 1 if the location i has the activity j assigned and 0 for other cases. It is equated as:

$$\min z = \sum_{i,j=1}^{n} n \sum_{h,k=1}^{n} d_{ih} f_{jk} X_{ij} X_{hk} \tag{9}$$

The QAP is a more general problem of TSP.

7 Conclusion

With the growing of soft computing field in various disciplines such as medicine, biology, mathematics, civil engineering, chemical engineering, and many more, the scientists should know about the capability of soft computing and how to use this for advancement of the society. It includes the tools and soft computing techniques for real-world problems which humans face in day-to-day life. Ant colony optimization has given the best results so far for NP-hard optimization problems. Though having some disadvantages of slow convergence, many new variations and hybrid systems of ACO have also came into existence and proved successful for solving NP-hard problem. It is efficiently applied to some problems discussed above such as travelling salesman problem, job-shop scheduling problem, and quadratic assignment problem. It can also be used to solve some dynamic applications, graph coloring, and constraint satisfaction.

References

1. L.A. Zadeh, Fuzzy logic, neural networks, and soft computing, in *Fuzzy Sets, Fuzzy Logic, and Fuzzy Systems: Selected Papers by Lotfi A Zadeh*, (1996), pp. 775–782
2. S. Bhattacharyya, U. Maulik, S. Bandyopadhyay, Soft computing and its applications, in *Data Mining: Concepts, Methodologies, Tools, and Applications*, (IGI Global, 2013), pp. 366–394
3. J.C. Bezdek, On the relationship between neural networks, pattern recognition and intelligence. Int. J. Approx. Reason. **6**, 85–107 (1992). https://doi.org/10.1016/0888613X(92)90013-P
4. S. Kumar, *Neural Networks: A Classroom Approach* (Tata McGraw-Hill, New Delhi, 2004)
5. S.K. Das, A. Kumar, B. Das, A.P. Burnwal, On soft computing techniques in various areas. Comput. Sci. Inf. Technol. (CS & IT) **3**, 59–68 (2013)
6. B. Prasad, Introduction to neuro-fuzzy systems, in *Advances in Soft Computing Series*, vol. 226, (Springer, Heidelberg, 2000)
7. O. Cord'on, R. Alcal'a, J. Alcal'a-Fernandez, I. Rojas, Genetic fuzzy systems: What's next? An introduction to the special section. IEEE Trans. Fuzzy Syst. **15**(4), 533–535 (2007)
8. A.K. Deb, Introduction to soft computing techniques: Artificial neural networks, fuzzy logic and genetic algorithms, in *Soft Computing in Textile Engineering*, (Woodhead Publishing, 2011), pp. 3–24
9. J.H. Holland, *Adaptation in Natural and Artificial Systems* (University of Michigan Press, Ann Arbor, 1975)
10. D.E. Goldberg, *Genetic Algorithms in Search, Optimization and Machine Learning* (Addison-Wesley, Reading, 1989)
11. B. Muller, J. Reinhardt, M. Strickland, *Neural Networks: An Introduction*, 2nd edn. (Springer-Verlag, Berlin, 1995)
12. A.U. Rehman, A. Khanum, A. Shaukat, Hybrid feature selection and tumor identification in brain MRI using swarm intelligence, in *Frontiers of Information Technology (FIT), 2013 11th International Conference on*, (IEEE, 2013)
13. W.S. McCulloch, W. Pitts, A logical calculus of the ideas immanent in nervous activity. Bull. Math. Biophys. **5**, 115–133 (1943)
14. D.O. Hebb, *The Organization of Behavior* (Wiley, New York, 1949)
15. R. Rosenblatt, *Principles of Neurodynamics* (Spartan Books, New York, 1962)

16. B. Widrow, M.E. Hoff, Adaptive switching circuits, in *IRE Eastern Electronic Show & Convention (WESCON1960), Convention Record*, vol. 4, (1960), pp. 96–104
17. M.L. Minsky, S. Papert, *Perceptrons* (MIT Press, Cambridge, MA, 1969)
18. G.J. Klir, B. Yuan, *Fuzzy Sets and Fuzzy Logic* (Prentice-Hall of India, New Delhi, 2007)
19. D. Driankov, H. Hellendoorn, M. Reinfrank, *An Introduction to Fuzzy Control* (Narosa Publishing House, New Delhi, 2001)
20. J. Timothy, *Ross, Fuzzy Logic with Engineering Applications* (Wiley India, New Delhi, 2007)
21. W. Pedrycz, *Fuzzy Control and Fuzzy Systems* (Overseas Press India, New Delhi, 2008)
22. I.P. Cabrera, P. Cordero, M. Ojeda-Aciego, Fuzzy logic, soft computing, and applications, in *International Work-Conference on Artificial Neural Networks*, (Springer, Berlin/Heidelberg, 2009, June), pp. 236–244
23. O. Castillo, P. Melin, J. Kacprzyk, W. Pedrycz, Soft computing for hybrid intelligent systems, in *Studies in Computational Intelligence*, vol. 154, (Springer, Heidelberg, 2008)
24. R. Langari, J. Yen, Fuzzy logic at the turn of the millennium. IEEE Trans. Fuzzy Syst. **9**(4), 481–482 (2001)
25. S.W. Kercel, Industrial applications of soft computing. IEEE Trans. Syst. Man Cybern. **36**(4), 450–452 (2006)
26. E. Herrera-Viedma, O. Cord'on, Special issue on soft computing applications to intelligent information retrieval on the internet. Int. J. Approx. Reason. **34**(2), 89–95 (2003)
27. M. Nachtegael, E. Kerre, S. Damas, D.V. der Weken, Special issue on recent advances in soft computing in image processing. Int. J. Approx. Reason. **50**(1), 1–2 (2009)
28. M. Jamshidi, N. Vadiee, T. Ross, Fuzzy logic and control: software and hardware applications, vol. 2 (Vol. 2). Pearson Education (1993)
29. V. Torra, V, Y. Narukawa, Special issue on soft computing methods in artifcial intelligence (2008)
30. T.S. Dillon, S.C. Shiu, S.K. Pal, Soft computing in case based reasoning. Appl. Intell. **21**(3), 231–232 (2004)
31. K. Egiazarian, A.E. Hassanien, Special issue on soft computing in multimedia processing. Informatica **29**(3), 251–253 (2005)
32. F. Chiclana, E. Herrera-Viedma, S. Alonso, F. Herrera, Special issue on fuzzy approaches in preference modelling, decision making and applications. Int. J. Uncertain. Fuzziness Knowlege-Based Syst. **16**(Suppl 2) (2008)
33. L. Ding, *A New Paradigm of Knowledge Engineering by Soft Computing* (WorldScientific, Singapore, 2001)
34. R.A. Aliev, B. Fazlollahi, R.R. Aliev, *Soft Computing and Its Applications in Business and Economics* (Springer, Heidelberg, 2004)
35. Z. Ma (ed.), *Soft Computing in Ontologies and Semantic Web* (Springer, Heidelberg, 2006)
36. J. Medina, Overcoming non-commutativity in multi-adjoint concept lattices, in *IWANN 2009, Part I. LNCS*, ed. by J. Cabestany et al., vol. 5517, (Springer, Heidelberg, 2009), pp. 278–285
37. F.T. Martins-Bede, L. Godo, S. Sandri, C.C. Freitas, L.V. Dutra, R.J.P.S. Guimaraes, R.S. Amaral, O.S. Carvalho, Classification of schistosomiasis prevalence using fuzzy case-based reasoning, in *IWANN 2009, Part I. LNCS*, ed. by J. Cabestany et al., vol. 5517, (Springer, Heidelberg, 2009), pp. 1053–1060
38. C. Lopez-Molina, E. Barrenechea, H. Bustince, P. Couto, B.D. Baets, J. Fern'andez, Edges detection based on gravitational forces, in *IWANN 2009, Part I. LNCS*, ed. by J. Cabestany et al., vol. 5517, (Springer, Heidelberg, 2009), pp. 1029–1036
39. B.P. Silalahi, N. Fathiah, P.T. Supriyo, Use of ant colony optimization algorithm for determining traveling salesman problem routes. Jurnal Matematika MANTIK **5**(2), 100–111 (2009)
40. D.M. Dorigo, M. Birattari, T. Stützle, Ant colony optimization. IEEE Comput. Intell. Mag., 28–39 (2006, November)
41. M. Dorigo, T. Stützle, *Ant Colony Optimization* (Massachusetts Institute of Technology, 2004)
42. A. Engelbrecht, *Fundamentals of Computational Swarm Intelligence* (Wiley, 2005), pp. 85–131

43. R.C. Eberhart, J. Kennedy, A new optimizer using particle swarm theory, in *Proceedings of the Sixth International Symposium on Micro Machine and Human Science*, (IEEE Service Center, Piscataway, 1995), pp. 39–43
44. T. Tützle, S. Linke, Experiments with variants of ant algorithms. Mathware Soft Comput. **9**(2–3), 193–207 (2002)
45. T. Stützle, H.H. Hoos, Improving the Ant system: A detailed report on the MAX-MIN Ant system, Technical report AIDA-96-12, FG Intellektik, FB Informatik, TU Darmstadt, Germany, 1996
46. T. Stützle, H.H. Hoos, MAX-MIN Ant system and local search for combinatorial optimization problems, in *Me-ta-Heuristics: Advances and Trends in Local Search Paradigms for Optimization*, ed. by S. Voss, S. Martello, I. Osman, C. Roucairol, (Kluwer Academic Publishers, Dordrecht, 1999), pp. 137–154
47. M. Dorigo, V. Maniezzo, A. Colorni, *Positive Feedback as a Search Strategy* (Milano, 1991)
48. M. Dorigo, L.M. Gambardella, Ant colony system: A cooperative learning approach to the traveling salesman problem. IEEE Trans. Evol. Comput. **1**(1), 53–66 (1997, April). https://doi.org/10.1109/4235.585892
49. W. Liu, S. Li, F. Zhao, A. Zheng, An ant colony optimization algorithm for the multiple traveling salesmen problem, in *2009 4th IEEE Conference on Industrial Electronics and Applications*, (Xi'an, 2009), pp. 1533–1537. https://doi.org/10.1109/ICIEA.2009.5138451
50. M. Dorigo, V. Maniezzo, A. Colorni, Ant system: Optimization by a colony of cooperating agents. IEEE Trans. Syst. Man Cybern. B **26**, 29–41 (1996)
51. T. Stützle, H.H. Hoos, Max-min ant system. Futur. Gener. Comput. Syst. **16**, 889–914 (2000)
52. Z.J. Zhang, Z.R. Feng, A novel max-min ant system algorithm for traveling salesman problem, in *Proceedings of Intelligent Computing and Intelligent Systems (ICIS 2009)*, (IEEE Press, Piscataway, 2009), pp. 508–511
53. B. Bullnheimer, R. Hartl, C. Strauss, A new rank-based version of the ant system: A computational study. Cent. Eur. J. Oper. Res. **7**, 25–38 (1999)
54. H.B. Mei, J. Wang, Z.H. Ren, An adaptive dynamic ant system based on acceleration for TSP, in *Proceedings of Computational Intelligence and Security (Piscataway)*, (IEEE Press, 2009), pp. 92–96
55. P. Guo, Z.J. Liu, Moderate ant system: An improved algorithm for solving TSP, in *Proceedings of Seventh International Conference on Natural Computation (Piscataway)*, (IEEE Press, 2011), pp. 1190–1196
56. Z.C.S.S. Hlaing, M.A. Khine, Solving traveling salesman problem by using improved ant colony optimization algorithm. Int. J. Inf. Educ. Technol. **1**, 404–409 (2011)
57. G. Dong, W.W. Guo, K. Tickle, Solving the traveling salesman problem using cooperative genetic ant systems. Expert Syst. Appl. **39**, 5006–5011 (2012)
58. A.Q. Ansari, Ibraheem, S. Katiyar, Comparison and analysis of solving travelling salesman problem using GA, ACO and hybrid of ACO with GA and CS, in *Proceedings of IEEE Workshop on Computational Intelligence: Theories, Applications and Future Directions (Piscataway)*, (IEEE Press, 2015), pp. 1–5
59. D. Weyland, R. Montemanni, L.M. Gambardella, An enhanced ant colony system for the probabilistic traveling salesman problem, in *International Conference on Bio-Inspired Models of Network, Information, and Computing Systems*, (Springer, Cham, 2012, December), pp. 237–249
60. L. Yabo, Z. Shikun, Z. Feng, An improved ants colony algorithm for NP-hard problem of travelling salesman, in *Joint International Conference on Pervasive Computing and the Networked World*, (Springer, Cham, 2013, December), pp. 432–440
61. T. Stützle, M. López-Ibáñez, M. Dorigo, A concise overview of applications of ant colony optimization, in *Wiley Encyclopedia of Operations Research and Management Science*, (Wiley, 2010)
62. T.S. Utzle, M. Dorigo, 1 ACO algorithms for the traveling salesman problemy

63. L.M. Gambardella, M. Dorigo, HAS-SOP: Hybrid ant system for the sequential ordering problem, Technical Report IDSIA 11–97, IDSIA, Lugano, Switzerland, 1997
64. V. Maniezzo, Exact and approximate nondeterministic tree-search procedures for the quadratic assignment problem, Technical Report CSR98–1, Scienze dell'Informazione, Universit∂a di Bologna, Sede di Cesena, 1998
65. T. St∫utzle, Local search algorithms for combinatorial problems | analysis, improvements, and new applications, PhD thesis, Darmstadt University of Technology, Department of Computer Science, 1998
66. T. St∫utzle, H.H. Hoos, MAX-MIN ant system and local search for combinatorial optimization problems, in *Meta-Heuristics: Advances and Trends in Local Search Paradigms for Optimization*, ed. by S. Voss, S. Martello, I. H. Os-man, C. Roucairol, (Kluwer, Boston, 1999), pp. 313–329
67. A. Puris, R. Bello, Y. Trujillo, A. Nowe, Y. Martínez, Two-stage ACO to solve the job shop scheduling problem, in *Iberoamerican Congress on Pattern Recognition*, (Springer, Berlin/Heidelberg, 2007, November), pp. 447–456
68. E. Flórez, W. Gómez, L. Bautista, An ant colony optimization algorithm for job shop scheduling problem. arXiv preprint arXiv:1309.5110 (2013)
69. S.A. Bidyarthy, G. Vivek, *Ant Colony Optimization for Quadratic Assignment Problem and School Bus Routing Problem* (Departement of Mathematics Indian Institute of Technology Guwahati, 2013)

Use of Kalman Filter and Its Variants in State Estimation: A Review

Neha Kumari, Rohan Kulkarni, Mohammed Riyaz Ahmed, and Naresh Kumar

1   Introduction

Electricity as a concept and a technology has rapidly improved the pace of human evolution. Once a luxury, it is now a necessity. With an increase in the population and thereby with rapid industrialization, effective management of electricity has become a major concern worldwide. The concept of the grid is one such solution toward the crisis. The increasing shift in the power sector toward renewable energy resources has made microgrids a necessity in the power system. These decentralized power stations with a definable boundary also act as a backup to the main power grid. Automated and computerized microgrids are smart grids that are intelligent, are reliable, and have highly efficient energy security. Further, conventionally used centralized grids are slower in decision-making, are inflexible, and have considerably high transmission losses with difficult troubleshooting processes. With ensured communication between the producers and the consumers, these issues are easily handled by microgrids. About 17% of the human population is deprived of the proper electrical facility. One of the main reasons for this is the lack of the required infrastructure in remote areas and an enormous amount of loss in energy during transmission and distribution. This could be avoided using microgrids which can be set up in distant places hence reducing the dependency on the main grid while

N. Kumari (✉) · R. Kulkarni
School of Electrical and Electronics Engineering, REVA University,
Bengaluru, Karnataka, India

M. R. Ahmed
School of Multidisciplinary Studies, REVA University, Bengaluru, Karnataka, India
e-mail: riyaz@reva.edu.in

N. Kumar
SCSE Galgotias University, Greater Noida, Uttar Pradesh, India
e-mail: kumar.naresh@galgotiasuniversity.edu.in

significantly decreasing the need for long power transmission. Also, it can contribute widely to a self-sufficient power generation. Since it uses generation methods locally, it emphasizes on renewable sources rather than non-renewable sources, hence contributing toward a greener environment.

Microgrids are a solution for providing energy to rural and remote areas; these vastly work with renewable sources of energy, hence are a way to green and clean generated power. With the progress in technology, there is an advancement in the use of renewable energy, making its application cheaper and hence making microgrids more economical. The setback in the usage of the utility grid is that it uses a variety of fuels to generate electricity, the growing future energy needs add to the increase in the power plants, leading to bulk usage of fossil fuels. These are creating a huge impact on the environment, affecting the health and quality of the ecosystem and causing climate changes. These can be overcome by the usage of renewable energy-dependent microgrids. About 1/5th of the world's electrical power production comes from renewable energy; this can be increased with more usage of clean energy microgrids. These reduce carbon emission while meeting sustainable development needs.

Integrating microgrids with main grids nationwide will create smart cities that are equipped with better energy management techniques. These will provide a way for efficient usage of resources, also real-time power supply needs can be met easily. As both smart cities and microgrids work on real-time data, we can manage energy resources easily. Internet of Things (IoT) is used as a means of communication, where sensors are deployed for collecting information that can be used for smart and efficient control of microgrids, also for better regulation of energy. Integrating smart cities and smart grids improves overall performance as well as quality and accuracy of the system, reducing wastage of energy and hence making a strong, reliable, and robust infrastructure.

State estimation, a data processing algorithm that predicts the response and does the real-time modelling of the power system plays a key role in efficient monitoring and control of a microgrid. The data assimilation for state estimation is realized by PMU (phasor measurement unit). This device congregates the data regarding electrical phasor quantities, such as voltage, frequency, and current in the electricity grid. PMUs are synchronized with global time stamps which allow synchronized real-time measurements of multiple remote points on the grid. When sufficient numbers of PMU's are deployed in the system, the system becomes completely observable. Smart metering devices and PMUs cannot be deployed due to high price and communication burden; this calls for state estimation to predict the response of the system. State estimation is required to estimate parameters like bus voltages, real and reactive power, transformer tap position, power dissipation, the scale of the grid, etc. SE works on recursive methods of prediction and estimation. SE provides some of the advantages like load forecasting, smoothing of errors, bad data detection, transient stability, security assessments, etc.

Since state estimation is essential for the monitoring and the supervisory control of the microgrid, it requires the estimation of the variables of the system. Artificial neural network (ANN), weighted least square (WLS), fuzzy logic, and H infinite are few of such ways. One among them is using Kalman filter. It usually is a linear quadratic estimator, which uses a series of measurements observed over time and produces estimates of the unknown variables. Further, it is an optimal state estimation technique for continuous and discrete systems which fuses various sensors and gives a better accurate result. The filter is not only used to find system states but is also used for system models. It uses mathematical equations that are used recursively to get more efficient estimates of the state. Defined generally by two steps, prediction and updating, the former uses the latest measurements to minimize the errors to refine the previous mathematical model. Minimizing the errors makes the estimated model very close to the real model, through which the required measurements can be obtained.

Kalman filter being a linear estimator cannot be used in all systems. Hence modified versions of Kalman filters are being used to identify system models and parameters. Extended (EKF), ensembled (EnKF), unscented (UKF), and particle Kalman filters are few of such filters. Modified versions of KF have been used to increase the accuracy of the estimation process. There are different methods for this filtering process. for example, EKF linearizes the system dynamics every step using the Jacobian matrices and then proceeds as for a linear system. UKF uses the probability distribution of states, EnKF uses probability density for much accurate estimation, and particle filter uses Gaussian distribution. These algorithms can be compared on a different basis such as accuracy, robustness against modelling and measurement errors, speed of computation, scalability, etc. Though KF is the most focused solution toward the microgrids in general, there have been several solutions aiming toward the same problem.

1. Fuzzy logic: Unlike Boolean logic, the fuzzy logic is utilized in situations where it is hard to describe the state in absolute 0 or 1. It provides n number of intermediate values in the range $\in [0,1]$. Thus, being more accurate and adding flexibility to the system.
2. Monte Carlo: Used to solve problems with probabilistic interpretation, Monte Carlo is a randomized algorithm which simulates the behavior of the systems. It uses randomness and statistics to get results.
3. WLS filter: This system makes the use of available data to draw a regression line. Here each data point is assigned a value based on its variance or standard deviation. The regression line fits best in available data such that the distance from the regression line and data points is minimum and can also be used to predict the values of the state.
4. Artificial neural network (ANN): ANN uses patterns and simple calculations to estimate the state of the system. Along with KF, it is used for the prediction of future estimates from the system using a large number of random noises sources [2]. The average estimates for all sets of noises are used as a training set for the neural network. The network provides a satisfactory estimate for the system when the next set of input noises are taken by the system. Hence it makes the estimation process faster and easier.

5. Belief propagation: Also known as the sum-product message passing algorithm, it calculates the marginal distribution of an unobservable node based on observable nodes. Marginal distribution is computed with the use of graphs like Markov's chains and Bayesian graphs. Further, by making an exact inference algorithm on trees by exploiting the poly tree structure, the propagation allows the margin to be computed much more efficiently.

6. Bayesian filter: Bayesian filter is based on the set of previous results and can also work with similar models like WLS or Monte Carlo that depend on the weight or probability of a condition. Along with some probability chains, it uses a Markov chain and trees to predict the most likely outcome. Besides quantizing all conditions of a system, it also uses prediction and updating hence works with other $\alpha-\beta$ filters like Kalman filter and can be used to enhance the performance of the other filters.

7. H ∞: H ∞ is a method that applies to a problem involving multivariate systems with cross-coupling between channels. It verifies the parameter uncertainty by designing an unbiased linear periodic asymptotical stable observer. [n] focuses on the particular problem of parameter identification using the H ∞ criterion. It is a method used for the formulation of problem optimization for more stabilization and guaranteed performance.

8. Rayleigh's filter: Rayleigh's is the mathematical model that is based on a noisy measurement process and has more accuracy of estimates and robustness. It treats the final measurement as a function of the state vector, H matrix, and the error signal. The only approximation introduced by the algorithm is to replace a conditional distribution to be matched to normal distribution at a signal point in each iteration which adds transparency to the assumption. This isolation of approximation allows for further analysis and potential improvement. In the comparison of Rayleigh's filter with other filters, it is found to be superior to other filters under various conditions [3].

9. Wieners filter: It is a noise filtering based Fourier iteration algorithm which is used to minimize the error between the actual output and the desired output to eliminate noises. The filter estimates the unknown signal by using a related signal as an input, filtering this unknown signal to produce the estimate of the output. Weiner follows a statistical approach for the estimation process; it has the main advantage of short computational time to find a solution.

2 Background

Maintaining a good power quality is one of the biggest challenges that microgrids are facing. Since no source is ideal, with the increase in the number of generators, i.e., by decentralization, the maintenance of constant voltage and frequency is a challenge. Power quality is mainly affected by the existence of transient voltages, irregular changes in load, harmonics, voltage sag, and swells. Maintaining the sinusoidal waveform could also be a challenge due to the same reasons. In simple words, power quality is the ability of the system (grid) to supply constant voltage and frequency with sinusoidal waveform; although it is impossible to eliminate the variations in power grids, they can, however, be maintained under tolerance limits. Constant monitoring and prediction of states in grids will be a boon to maintaining good power quality. Power quality can also be improved using PWM active filters, power conditioning equipment like passive shunt L-C Filters, and correcting power factor. Tuomo et al. use the power consumption model for analyzing the power supply behavior and for improving the energy efficiency by using KF [4]. The work also proposes the use of multiple models and parameters identifying KF for verification and further improvement of system models for power supply and related systems. Sharma et al. test and validate multiple power system dynamic state estimation algorithms using real-time digital simulator (RTDS) by employing EKF, UKF, and CKF approach [5]. Simulation results on two test systems are used to get the relative performances of the three filters. The result of the simulation shows that CKF is the fastest, most accurate, and stable filter under sudden load change conditions compared to the other two filters. Rana et al. propose the use of adaptive-then combine distributed dynamic approach to monitor the varying wind power generation pattern under lossy communication links between wind turbines and energy management systems [6]. Sensors and the MSE are deployed to obtain the system state information and the local state information, respectively. A brief idea about the need of SE for a dynamic system is given by Zhou et al. by providing an overview on the different SE techniques such as EnKF, EKF, UKF, and PF [7]. Giving comparisons between SE filters using simulations of multiple scenarios based on MSEs. Investigations on the future development of the dynamic state, estimation has also been carried out by Rana et al. [8]. The problems on the modelling of the system dynamics and the tracking of the changes in the system in real-time, nodal load forecasting technique, to the hierarchical EKF, and the other replacing KF by static estimation technique. The work also discusses the advantages of using a static approach to state estimation. The capability of accurately tracking single phase harmonic phases when subjected to the varying nominal frequency and out of band inter-harmonic inference is discussed by Chakir et al. [9] which is then provided to phasor measurement units (PMUs) by extending synchrophasor algorithms. Two schemes are proposed for this application, the fastest solution is through KF bank and most accurate is through impulse response filter. The two ways are compared in terms of computational speed and performance using a four-cycle short-time fast Fourier transform algorithm. Yang et al. focus on the problem of inaccurate estimation of phase synchronization faced by large-scale deployment of PMUs [10]. A model for power system state estimation with phase mismatch is introduced, and alternating minimization and parallel Kalman filtering is used to demonstrate the improved accuracy of the algorithm. Rana et al. develop a KF-based optimal feedback control method for microgrid SE and stabilization [11]. This is done by taking system information from the noisy measurement and uncertainties using KF to estimate the system state, which is used for the application of control algorithms. The main limitations of the work are the execution time due to encoder and decoder and are overcome by data compression and model reduction method. Nayak et al. use

UKF for the dynamic tracking of harmonics in microgrid which gets induced due to the application of nonlinear loads [12]. A microgrid model and the parameter of UKF are selected and the outcomes of the application of the algorithm on the microgrid are studied in MATLAB/SIMULINK environment. An independent local KF which can be used for SE without communication is designed by Liu et al. [13]. The synthesized local KF is applied to the interconnected power networks, and the improved results compared to the previously proposed modelling approach are shown through the simulation results. For the voltage regulation control in the DER's at the distribution level, state information is collected from the smart sensors is communicated through a 5G communication network by Rana et al. [14]. Least square-based KF for SE is proposed for the feedback control method of voltage regulation. The performance index of this is optimized using a semi-definite programming technique. The prediction and approximation of future power demand are done by Farmer et al. using interacting multiple models KF's which is effective for year-round perfection having high efficiency and calculations and less requirement of computational resources [15]. A brief overview of the industrial applications and implementation issues of KF in six topics of the industrial electronic community is presented by Auger et al. [16]. The application of state and parameter estimation for control and diagnosis in distributed energy resources is also discussed. A decentralized KF is designed for local SE in a distributed generation-based power system avoiding the use of communication channels by Liu et al. [17]. The simulated results of the DKF approach are also discussed. The comparison between different algorithms, namely, KF, EKF, and adaptive EKF, based on the accuracy of computing practical harmonics is carried out by Nayak et al. [18]. Using a microgrid model, the algorithm's accuracy is determined through MATLAB/SIMULINK. The need for applying IoT to the microgrid state-space model after linearizing about a point so that the SE of the DER can be done using the 5G network is discussed by Rana et al. [19]. The simulation results show the execution of the proposed method of DERSE through IoT with a 5G network. They also propose KF-based microgrid SE through IoT communication network under two different sensing scenarios, for real-time monitoring of the power system and also for the customer utilization of the power grid [20]. Simulation results show the ability of the KF-based microgrid SE to estimate the system state in all scenarios. The estimation of harmonics is important for the power quality estimation in the power system for the accurate estimation of amplitude, phase, and frequency. Nayak et al. present a new filter, namely, adaptive robust KF (REKF) based on a linear H infinity technique for estimating the fundamental and harmonic phasors and frequency in the DG system for islanded, non-islanded, and switching events [21]. A distributive system state estimation that detects a change in the topology of the model through recursive Bayesian estimator, which uses forecast-aided state estimation (FASE) which is based on Monte Carlo simulations, is proposed by Huang et al. [22]. A comparison between WLS, EKF, and UKF is done where UKF outperforms the others with the least variance even when the topology is unknown. A model is predicted by the use of single-shot WLS which performs very well especially in cases where model changes are slow. Rana et al., in their other work focus on efficient communication infrastructure to facilitate estimation of distributed energy resources state of DER, is treated Dynamic and SE approach is based on recursive systematic convolutional code [23]. Use of DKF for weight averaging method for decentralization and UKF for real-time power system SE. Further, they aim at achieving accurate SE and wide-area real-time monitoring of the intermittent energy sources through discrete-time linear quadratic regulation control and KF-based SE [24]. The simulation results show that the proposed method outperforms the traditional methods used for SE. These results of accuracy-dependent KF and DLQR schemes are very promising for future applications in smart grids. State estimation using WLS and EKF as an alternator method has been discussed by Shabani et al. [25]. The work validates that EKF outperforms WLS and hence proposes EKF as an alternator method for SE over the traditional WLS methods. Masud et al. proposes a two-way communication from microgrids using IoT state information collection and estimation and addresses the challenge of maintaining voltage regulation through deployment and control of sensors and actuators [26]. The proposed KF works efficiently even when there is a large number of sensors in the system. State estimation generated through the ladder iterative technique is looked into by Ratmir et al. [27]. It compares through ladder iterative technique and WLS and EKF to find the most probable state based on the available measurements. It is shown that KF requires fewer iterations, but it also needs some foreknowledge about the states. Yang et al. address the problem of mismatch in time-synchronized sensors, the authors also propose a new measurement model for SE based on posterior Cramer Rao bound [28]. Baldi et al. carry out their work on fault detection and diagnosis tools [29]. They use duel estimation schemes based on EKF and UKF for building heat transfer models to cope up with the low sampling rate of data. Ying et al. propose a novel approach to construct a stochastic model where microgrids are modelled as a factor graph [30]. They use belief propagation to draw statistical inference on the factor graph using belief propagation to monitor un-metered node behavior (Table 1).

3   Mathematical Foundation of Kalman Filter

In order to learn a lower dimensional depiction, GNE combines gene interaction network and gene expression data [24]. Now, first we go with recommending a data pipeline using various semantic and machine learning (ML) methods to predict novel ontology-dependent annotations of activated genes; then we implement a new semantic priority instruction to compartmentalize expected annotations by their probability of being accurate. Our experiments and validations proved the efficacy and importance of the expected annotations in our pipeline, by choosing as most likely several forecasted annotations that were later validated. The nodes are genes, so the same-colored genes have identical vocal profiles. GNE groups genes with identical network topology that are related in the graph or have a common neighborhood and allocate similarities (common profiles of articulation) in the implanted space which is illustrated using.

Table 1 Summary of all Kalman-based state estimators

State estimator	System model	Assumed distribution	Computational cost	Remarks
Kalman Filter (KF)	Linear	Gaussian	Low	Doesn't handle very complex systems with a greater number of variables
Extended KF	Locally linear	Gaussian	Low	Not suitable for larger power systems and due to linearization divergence may occur
Unscented KF	Nonlinear	Gaussian	Medium	Performance deteriorates with an increase in the number of state variables
Ensembled KF	Nonlinear	Non-Gaussian	High	Suitable for strong nonlinear systems and measurement interpolation doesn't affect its performance
Particle KF	Nonlinear	Non-Gaussian	High	Suitable for more complex system models which are not well defined, not robust to outliers

Kalman Filter [KF], a recursive algorithm that works on dynamic systems and uses measurement, estimation, prediction, and updating were developed by Rudolf E Kalman. It is a linear quadratic estimator and hence works on a discrete linear system of the form:

$$xk = Ak\,xx - 1 + Bk\,uk + wk \tag{1}$$

with $x_k \in R^n$, A_k, w_k, and B_k representing transition matrix, process noise, and control input model, respectively, and measurement (or observation) of z_k is:

$$z_k = H_k x_k + v_k \tag{2}$$

with $z_k \in R^m$ and H_k being observation model if process noise w_k and the measurement noise v_k are assumed to be an independent random variable with Gaussian probability density functions and zero mean value. Let's go about to understand the model with equations, first of which is the state update equation.

$$\hat{x}_k = \hat{x}_{k-1} + K_k \left(z_k - H_k \hat{x}_{k-1} \right) \tag{3}$$

here x_{k-1} represents the previous estimate for the kth instant. The difference between the latest measurement and the previous estimate gives the measurement residual called innovation represented by y_k. The weight of innovation decreases with a greater number of iterations as the Kalman gain decreases. This state update equation gives the present estimate at the kth instant. State extrapolation equation is the immediate shift from the state update equation after an instant. This equation extrapolates the current state to the next state,

$$\hat{X}_{k-1} = \hat{X}_{k-1|k-1} + K_k y_k \tag{4}$$

hence giving the prediction of the next state. From [3], the factor that is responsible for the addition of the difference between the present measurement and the previous estimate to the previous estimate is called Kalman gain. This is defined by the third Kalman filter equation, Kalman gain equation:

$$K_k = P_{k-1} H_k^T \left(H_k P_{k-1} H_k^T + R_K \right)^{-} \tag{5}$$

Which is the ratio of the in estimate to the sum of the uncertainty in estimate with the uncertainty in the measurement?

$$\left[0 \le K_n \le 1 \right] \tag{6}$$

From [3],

$$\hat{x}_k = \hat{x}_{k-1} \left(1 - H_k K_k \right) + z_k K_k \tag{7}$$

hence, the weight K_k is given to the measurement and $1 - HK_k$ is given to the estimate, as the number of iterations increases the significance of the measurements decreases and the new value comes closer to the previous estimate value. The uncertainty in the estimate needs to be updated with the updating of the state of the dynamic system, for finding the present uncertainty in the estimate using the previous uncertainty, estimate uncertainty update equation, or covariance update equation,

$$P_k = \left(I - K_k H_k \right) P_{k-1} \tag{8}$$

Where

$$\left(1 - K_k \le 1 \right) \tag{9}$$

As the number of iterations increases, uncertainty in the present estimate decreases. Estimate uncertainty extrapolation equation is applied to realize the present uncertainty using the previous uncertainty.

$$P_{k-1} = A_k P_{k-1|k-1} A_k^T + Q_k \tag{10}$$

So, [4, 10] is categorized in priori, and represents the prediction of co-variance and state of the system [3, 8] is categorized in posteriori and represents the correction of covariance and state of the system (Fig. 1).

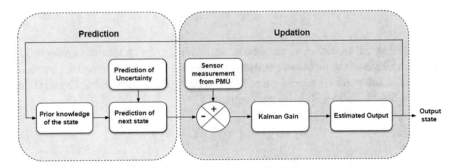

Fig. 1 A typical representation of an RFID system in a manufacturing industry installed at the packing and distribution unit

3.1 Extended Kalman Filter

Kalman filter works on a linear model since most of the systems are nonlinear; the filter is reformed to extended Kalman filter, hence the name [1]. The filter works on nonlinear systems, represented as follows:

$$x_k = f\left(x_{k-1}, u_k + w_k\right) \tag{11}$$

$$z_k = h\left(x_k\right) + v_k \tag{12}$$

where w_k and v_k represent the process and observation noise, respectively, where both are assumed to be mean multivariate Gaussian noises, with covariant Q_k and R_k, respectively. Here in this equation, Uk represents a control vector. Predicted state and measurement can be computed using the nonlinear function f and h from the previous estimate and the control vector and predicted state, respectively. The matrix of the partial derivative of f and h is computed and then is further used in the equation of the extended Kalman filter. The finding of the Jacobin of the nonlinear function tends to linearize the dynamic system around the current estimate at every step. EKF approximately linearizes the nonlinear model about the current estimate and then uses linear Kalman filter for further prediction of the estimate.

Predict

$$\hat{x}_k = f\left(\hat{x}_{k-1}, u_k\right) \tag{13}$$

$$\hat{P}_k = F_k \hat{P}_{k-1} + F_k^T + Q_k \tag{14}$$

Update

$$K_k = P_{k-1} H_k^T \left(H_k P_{k-1} H_k^T + R_k\right)^{-1} \tag{15}$$

$$\hat{x}_k = \hat{x}_{k-1} + K_x \hat{y}_k \tag{16}$$

$$P_k = \left(I - K_k H_k\right) P_{k-1} \tag{17}$$

As the number of iterations of the EKF increase, the linearity improves hence decreasing the linearization error by repeated modification at the center point of first order Taylor's series expansion, at the current estimate. EKF is a non-optimal estimator as both the state transition model and the measurement are nonlinear. Due to linearization, inaccuracies in the initial state estimation or process model may lead to the divergence of the filter from the true value.

3.2 Unscented Kalman Filter

The performance of the UKF exceeds that of EKF as it uses multiple points called weighted sigma points, instead of a single point, hence given improved approximation. UKF falls under the category of Sigma point Kalman Filter or linear regression Kalman filter. The points in the present Gaussian are mapped in the predicted Gaussian through nonlinear function, by which new mean and variance of the predicted Gaussian can be calculated. UKF involves the process of unscented transform, computing the set of sigma points and assigning the weight of each sigma point and transferring these points through nonlinear function, computing the predicted Gaussian and hence finding the predicted mean and variance. The number of sigma points depends on the dimensionality of the system, 2 N + 1, where N denotes dimensionality.

$$X^{[0]} = \mu \tag{18}$$

$$X[i] = \mu + \left(\sqrt{n + \lambda \Sigma}\right) n - i \quad , for\ i = 1, \ldots, n \tag{19}$$

$$X^{[i]} = \mu - \left(\sqrt{n + \lambda}\ \Sigma\right)_{n-i} \mu - \left(\sqrt{n + \lambda \Sigma}\right) n - i \quad , for\ i = n + 1, \ldots, 2n \tag{20}$$

X: Sigma point matrix.
μ: Mean of the Gaussian.
n: Dimensionality of the system.
λ: The scaling factor which tells of how far a sigma point should be from the mean point.
P: Covariance matrix weights of the sigma points are calculated as follows:

$$W[0] = \frac{\lambda}{n + \lambda} \tag{21}$$

$$W[i] = \frac{1}{2(n + \lambda)} w^{[i]} = \frac{1}{2(n + \lambda)} for\ (i = 1, 2 \ldots 2n) \tag{22}$$

The sum of all the weights is equal to 1. The prediction of mean and covariance is done after considering the process noise.

$$\mu' = \sum_{i=0}^{2n} w^{[i]} g\left(X^{[i]}\right) \tag{23}$$

$$\sum{}' = \sum_{i=0}^{n} w^{[i]} \left(g\left(X^{[i]}\right) - \mu'\right) + \left(g\left(X^{[i]}\right) - \mu'\right)^{T} \tag{24}$$

μ': predicted mean
$\sum'$: predicted co-variance
w: weights of the sigma point
g: nonlinear function

Now, the updation step is computed after getting the measurements from the sensors.

$$Z = h(x) \tag{25}$$

$$\hat{z} = \sum_{i-0}^{2n} w^{[i]} z^{[i]} \tag{26}$$

$$S = \sum_{i=0}^{2n} \left(Z^{[i]} - \hat{z}\right)\left(Z^{[i]} - \hat{z}\right)^{T} + Q \tag{27}$$

Z: transformed sigma points in measurements space
$\hat{z}$: mean in measurement space S – co-variance in measurement space
Q: noise
H: function that maps sigma points to measurement space

The association between the sigma points in state space and measurement space is given by

$$T = \sum_{i=0}^{2n} w^{[i]} \left(X^{[i]} - \mu'\right)\left(Z^{[i]} - \hat{z}\right)^{T} \tag{28}$$

$$K = TS^{-1} \tag{29}$$

T: cross-co-relation matrix between state space and predicted space
S: predicted co variance matrix
K: Kalman gain

$$\mu = \mu' + K\left(Z - \hat{z}\right) \tag{30}$$

$$\sum{} = \left(I - KI\right)\sum{}' \tag{31}$$

3.3 Ensemble Kalman Filter

EnKF is an alternative to traditional EKF as it is not feasible to use EKF for linearizing strongly nonlinear dynamics. It is an implementation of the Bayesian update using Monte Carlo and works on a huge number of variables. It performs sequential data assimilation for ensembled forecasting and is used for the advancement of the covariance matrix. It can handle computationally complex systems with a greater number of variables; KF is not feasible for such high dimensional systems. It discretizes the system state into a collection of state vectors, each called ensemble. Here probability density function is updated by advancing each member of the ensemble which represents the new state analyzed. By integrating the ensemble of the state model, the mean and error covariance can be found. As the number of ensembles increases the probability density function can be approximated more accurately. It reduces the computational work, as it is sufficient to use a limited number of model states for reasonable statistical convergence. $X_2[x_1,....x_n] = [x_i]$ is n × N matrix, and prior ensemble matrix, $D = [d_1,......d_n]$, is m × N matrix, where d is the data column and has random vector from n-dimensional normal distribution N(0,R), and the column is given by

$$X^\wedge = X + K\left(D - HX\right) \tag{32}$$

which gives a random sample from the posterior distribution. We can get EnKF by replacing Q from Kalman gain matrix,

$$K = QH^T \left(HQH^T + R\right)^{-1} \tag{33}$$

Co-variance C computed by using ensemble members

$$E(x) : \frac{1}{N}\sum_{k=1}^{N} x_k \tag{34}$$

$$C = \frac{AA^T}{N-1} \tag{35}$$

where E(x), C gives ensemble mean, co-variance, and

$$A = X - E(x) \tag{36}$$

Hence, the posterior ensemble is given by

$$X^P = X + (H^T \left(H\left(H^T + R\right)^{-1}\left(D - HX\right)\right) \tag{37}$$

3.4   Particle Kalman Filter

Implementation of the Monte Carlo technique is more optimal for the nonlinear or
non-Gaussian distribution systems. These offer a more robust approach; this methodology is used by particle Kalman filter [PKF]. These use several numbers of random samples (or particles) that have their weights that are used to find the posterior
density function at any state-space model. Hence, these are used to compute the
estimates using the weighted samples. PKF does not use linearization technique like
EKF and is computationally expensive. These methods do not perform well with
high dimensionality systems. As the number of samples becomes very large, Monte
Carlo characterization becomes an equivalent representation to the usual functional
description of the posterior probability density function, hence the sequential importance sampling filter [SIS] approaches optimal Bayesian filter. The weighted ran-
dom samples used for the description of probability distribution function $P(x_{0:k}|Z_{1:k})$
is represented by

$$\left\{x_{0:k}^i, w_k^i\right\} \tag{38}$$

where

$$\left\{x_{0:k}^i, i = 0,\ldots,n\right\} \tag{39}$$

is a set of samples and

$$\left\{w_k^i, i = 1,\ldots N_s\right\} \tag{40}$$

is the weight associated with the samples?

$$P\left(x_{0:k}|Z_{1:k}\right) \approx \sum_{i=1}^{N_s} W_k^i \delta\left(x_{0:k} - X_{0:k}^i\right) \tag{41}$$

Here the weights are taken by using importance sampling. The samplings (support
points) are taken sequentially as measurements, and in this SIS algorithm, the
weights are updated to approach the true posterior density.

3.5   Cubature Kalman Filter

Cubature Kalman filter is the closest known approximation to the Bayesian filter
that can be used for the nonlinear systems. EKF is not the best choice for many
applications as it is only for mild nonlinear systems. CKF employs deterministic
sampling to evaluate the interactive integrals encountered in filtering problems, it is
used in nonlinear state estimation problems and is more accurate and stable

compared to UKF. Cubature shows more improvement in error performance for higher degree cubature's and shows more accuracy and stability at sudden load change.

3.6 Further Types of Kalman Filter

Parallel Kalman filter is a type of KF where measurement sequence is divided into smaller segments for parallel processing due to large amounts of data for fast and accurate real-time computing. Decentralized Kalman filter, this is designed for a network of sensors with central Kalman filter. It aims at minimizing the computational load on the system. Adaptive Kalman filter, these are designed to adapt to the changes in the system model autonomously. It uses various estimators like adaptive interacting multiple model (ADIMM). Complex Kalman filter, this filter exhibits very little oscillation in frequency during the changes in the state of the model. Fuzzy Kalman filter, it is developed by incorporating interval Kalman filter (which only works at necessary interval arithmetic). Similar to fuzzy logic, it is based on the IF-THEN rule. It fuzziness the internal system matrices.

4 Discussion

Microgrids, stand-alone distributed systems are required to shift the present power system toward renewable sources of energy. IoT is used to communicate the information retrieved from sensors and is used in smart grids for smart connectivity. Its use in smart grids for energy management systems extends the benefits of a smart grid beyond automation, distribution, and monitoring. Smart grids require state estimation for real-time supervisory control and monitoring. There are various techniques to perform state estimation, some of which are discussed in this paper.

This paper on the whole deals with the Kalman filter, which is a recursive linear estimator that fuses the multiple numbers of sensors, used for the system state estimation. Here variants of Kalman filter and their differences and improvements are discussed. KF works on linear systems, hence creating a need for the development of the filter to be suitable for nonlinear systems. The variants can be compared based on different parameters which include accuracy, robustness against modelling error and measurement error, speed of computation, and scalability. It is seen that EnKF outperforms other filters when a typical PMU sampling rate is used for estimation. CKF is the closest known approximation to the Bayesian filter which work so nonlinear systems [5]. It's also seen that measurement interpolation improves the accuracy of EKF, UKF, and PKF and does not show any significant influence on the performance of EnKF [7]. Also, a greater number of samples improve the estimation accuracy and convergence of PF. Taking robustness against modelling of errors and measurements into account, all the filters are affected by the outliers but they soon converge back to the true value except for PF, which does not regain accuracy quickly after the anomaly is removed. Though PKF is theoretically superior compared to other filters, it has more sensitivity toward errors in the measurements and is also more complex computationally; it is inefficient compared to EnKF, which is more robust to outliers and missing data and which reduces the computational burden, hence making it one of the most optimal state estimation filter. It is identified that EnKF is computationally feasible for high dimensional systems and shows properties of greater robustness, accuracy, and potential scalability. Here, different literature on the implementation of the Kalman Filter has been presented.

5 Conclusion

Zero carbon energy is said to be the only sustainable way to meet the energy hunger of the planet. One of the most discussed and researched area is stand-alone distributed system microgrids that are currently seeing an appreciable adoption worldwide. The decentralization of the conventional power grid not only improves the power management but also greatly reduces transmission losses and troubleshooting. With an expected increase in the power generation and distribution, the effective management of these microgrids contributes to a stable, smart, and cost-effective energy system. Stand-alone DC microgrids are playing a crucial role in the evolution of power generation and transmission. They are not only an improvement to conventional grids but are being a complement to the existing conventional system by reaching the remote places. The increased decentralization of power generation demands smart grids which in turn require efficient ways to monitor the grid and estimation of its state and models.

Among various available state estimation techniques, Kalman filters provide a better optimization of the state parameters. The state estimation by Kalman filter enables us to load forecasting and bad data detection and to attain efficient energy security. This paper investigates the various ways of distributed system state estimation via Kalman filters and its variants. The work focuses in detail the working of Kalman filter with different works related to it discussed briefly. Comparisons on the variants of Kalman filter have been carried out and optimal filters for various settings are being advised based on their discussed mathematical modeling. Though there is good literature in this regard, the pursuit of an efficient filter to enhance monitoring and state estimation in microgrids has just begun. The most suitable filter for islanded mode and grid-connected mode with a 5G ecosystem for IoT devices for efficient monitoring is yet to be found. However, the distributed system SE by Kalman filter is not only appreciated but also recommended with its variants to be used under different parameter environments. With more optimization of the performing index, and appreciable flexibility factor by variants, Kalman filter are hence recommended to improve the state estimation of the distributed microgrids.

Acknowledgments We are indebted to Abhinandan A. J, Bhoomika C. M, and Anjanakumari B. T. of School of ECE, REVA University, Bangalore, for their valuable inputs and enduring support during the work. We would also like to express our deep gratitude to REVA University for their unconditional support in carrying this work.

References

1. P. Rousseaux, T. Van Cutsem, T.E.D. Liacco, Whither dynamic state estimation? Int. J. Electr. Power Energy Syst. **12**(2), 104–116 (1990)
2. A.J. Kanekar, A. Feliachi, State estimation using artificial neural networks, in *[1990] Proceedings. The Twenty-Second Southeastern Symposium on System Theory*, (IEEE Computer Society, 1990), pp. 552–553
3. J.M.C. Clark, R.B. Vinter, M.M. Yaqoob, Shiftedrayleigh filter: A new algorithm for bearings-only tracking. IEEE Trans. Aerosp. Electron. Syst. **43**(4), 1373–1384 (2007)
4. T. Malkamki, S.J. Ovaska, Optimal state estimation for improved power measurements and model verification: Theory, in *2011 International Green Computing Conference and Workshops*, (IEEE, 2011), pp. 1–6
5. A. Sharma, S.C. Srivastava, S. Chakrabarti, Testing and validation of power system dynamic state estimators using real time digital simulator (RTDS). IEEE Trans. Power Syst. **31**(3), 2338–2347 (2015)
6. M.M. Rana, L. Li, S.W. Su, An adaptive-then-combine dynamic state estimation considering renewable generations in smart grids. IEEE J. Sel. Areas Commun. **34**(12), 3954–3961 (2016)
7. N. Zhou et al., *Capturing Dynamics in the Power Grid: Formulation of Dynamic State Estimation Through Data Assimilation. No. PNNL-23213* (Pacific Northwest National Lab. (PNNL), Richland, 2014)
8. M.M. Rana, L. Li, S.W. Su, Distributed dynamic state estimation over a lossy communication network with an application to smart grids, in *2016 IEEE 55th Conference on Decision and Control (CDC)*, (IEEE, 2016), pp. 1–6
9. M. Chakir, I. Kamwa, H. Le Huy, Extended C37. 118.1 PMU algorithms for joint tracking of fundamental and harmonic phasors in stressed power systems and microgrids. IEEE Trans. Power Delivery **29**(3), 1465–1480 (2014)
10. P. Yang et al., Power system state estimation using PMUs with imperfect synchronization. IEEE Trans. Power Syst. **28**(4), 4162–4172 (2013)
11. M.M. Rana, L. Li, S.W. Su, Microgrid state estimation and control using Kalman filter and semidefinite programming technique. Int. Energy J. **16**(2), 47 (2016)
12. P. Nayak, S. Jena, Estimation of harmonics in microgrid using unscented Kalman filter. Int. J. Pure Appl. Math. **114**, 73–81 (2017)
13. J. Liu et al., State estimation and branch current learning using independent local Kalman filter with virtual disturbance model. IEEE Trans. Instrum. Meas. **60**(9), 30263034 (2011)
14. M. Rana, L. Li, S. Su, Kalman filter based microgrid state estimation and control using the IoT with 5G networks, in *2015 IEEE PES Asia-Pacific Power and Energy Engineering Conference (APPEEC)*, (IEEE, 2015), pp. 1–5
15. M.E. Farmer, M. Allison, Power demand prediction in smart microgrids using interacting multiple model Kalman filtering, in *Proceedings of the Workshop on Communications, Computation and Control for Resilient Smart Energy Systems*, (2016), pp. 1–8
16. F. Auger et al., Industrial applications of the Kalman filter: A review. IEEE Trans. Ind. Electron. **60**(12), 54585471 (2013)
17. J. Liu et al., State estimation and learning of unknown branch current flows using decentralized Kalman filter with virtual disturbance model, in *2010 IEEE International Workshop on Applied Measurements for Power Systems*, (IEEE Computer Society, 2010), pp. 552–553

18. P. Nayak et al., Comparative study of harmonics estimation in micro grid using adaptive extended Kalman filter, in *2016 International Conference on Electrical, Electronics, and Optimization Techniques (ICEEOT)*, (IEEE, 2016), pp. 434–438

19. M. Rana, L. Li, S. Su, Kalman filter based distributed state estimation with communication systems, in *2015 IEEE PES Asia-Pacific Power and Energy Engineering Conference (APPEEC)*, (IEEE, 2015), pp. 1–5

20. M.M. Rana, L. Li, Kalman filter based microgrid state estimation using the internet of things communication network, in *2015 12th International Conference on Information Technology-New Generations*, (IEEE, 2015), pp. 501–505

21. P. Nayak, B.N. Sahu, A robust extended Kalman filter for the estimation of time varying power system harmonics in noise, in *2015 IEEE Power, Communication and Information Technology Conference (PCITC)*, (IEEE, 2015), pp. 635–640

22. J. Huang, V. Gupta, Y.-F. Huang, Electric grid state estimators for distribution systems with microgrids, in *2012 46th Annual Conference on Information Sciences and Systems (CISS)*, (IEEE, 2012), pp. 1–6

23. M. Rana, L. Li, S. Su, Distributed microgrid state estimation using smart grid communications, in *2015 IEEE PES Asia-Pacific Power and Energy Engineering Conference (APPEEC)*, (IEEE, 2015), pp. 1–5

24. M. Rana, L. Li, An overview of distributed microgrid state estimation and control for smart grids. Sensors **15**(2), 4302–4325 (2015)

25. F. Shabani et al., State estimation of a distribution system using WLS and EKF techniques, in *2015 IEEE International Conference on Information Reuse and Integration*, (IEEE, 2015), pp. 609–613

26. M.M. Rana, L. Li, Renewable microgrid state estimation using the internet of things communication network, in *2017 19th International Conference on Advanced Communication Technology (ICACT)*, (IEEE, 2017), pp. 823–829

27. R. Gelagaev, P. Vermeyen, J. Driesen, State estimation in distribution grids, in *2008 13th International Conference on Harmonicsand Quality of Power*, (IEEE, 2008), pp. 1–6

28. P. Yang, *Power System State Estimation and Renewable Energy Optimization in Smart Grids* (Washington University, Saint Louis, 2014)

29. S. Baldi et al., Dual estimation: Constructing building energy models from data sampled at low rate. Appl. Energy **169**, 8192 (2016)

30. Y. Hu et al., Micro-grid state estimation using belief propagation on factor graphs. *Asia-Pacific Signal and Information Processing Association Annual Summit and Conference* (2010)

Precoder and Combiner Optimization in mmWave Hybrid Beamforming Systems

Abdul Haq Nalband, Mrinal Sarvagya, and Mohammed Riyaz Ahmed

1 Introduction

A large number of antennas will be used in the systems to defeat the severe path loss in mmWave frequency band [1]. The high power consumption and hardware cost of these large antenna systems restrict the use of conventional MIMO baseband precoding methods which dedicate a separate RF chain for each antenna element [2]. One way to overcome these constraints is to divide the needed precoding processing between analog and digital domains by designing hybrid analog-digital precoding schemes. A hybrid precoder and combiner arrangement exercised in mmWave MIMO systems is depicted in Fig. 1. The RF precoder/combiner sections are comprised of analog phase shifters with fixed amplitudes, and processes of baseband precoding/combining are achieved in the digital domain [3]. The transmitter having N_t antennas sends N_s data streams to user equipment with Nr antennas. To render multi-stream communication, transmitter and receiver systems comprise N_t^{RF} and N_r^{RF} RF chains, respectively. It is important to satisfy the conditions $N_s '' N_t^{RF} '' N_t$ and $N_s '' N_t^{RF} '' N_t$ [4]. The hybrid precoder $F \in \mathbb{C}^{N_t \times N_s}$ at the transmitter comprises a large dimensional analog precoder $F_{RF} \in \mathbb{C}^{N_t \times N_t^{RF}}$ and a small dimensional digital precoder $F_{BB} \in \mathbb{C}^{N_t^{RF} \times N_s}$, where C represents the complex numbers [5]. Similarly at the receiver, the combiner comprises RF combiner W_{RF} and baseband combiner W_{BB} with dimension $N_r \times N_r^{RF}$ and $N_r^{RF} \times N_s$, respectively.

A. H. Nalband (✉) · M. R. Ahmed
School of Multidisciplinary Studies, REVA University, Bangalore, Karnataka, India
e-mail: abdulhaq@reva.edu.in; riyaz@reva.edu.in

M. Sarvagya
School Electronics and Communication Engineering, REVA University,
Bangalore, Karnataka, India
e-mail: mrinalsarvagya@reva.edu.in

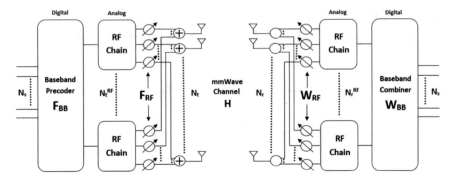

Fig. 1 Hybrid beamforming mmWave communication system

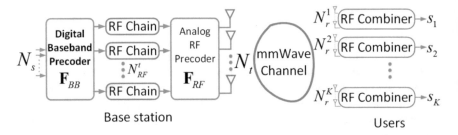

Fig. 2 mmWave Multiuser Downlink System

1.1 Multi-user MIMO Downlink System Model

Hybrid beamforming (HBF) is an assuring solution, especially when extended to a multi-user situation. Consider a mmWave multi-user MIMO downlink system as shown in Fig. 2.

The base station acts as a transmitter sending multiple data streams to each user equipment simultaneously and serving K user equipments (UEs) or mobile stations (MSs) at the same time [6]. Each UE k consists N_r^k antenna elements. We consider that the BS is equipped with K RF chains and capable to transmit K streams [7, 8]. When BS transmits one stream to each user, then it can serve K users simultaneously. The BS is assumed to have perfect knowledge of the channel state information, which can be achieved as UEs send the instantaneous ideal channel state information (CSI) information for the BS via a feedback mechanism as illustrated in Fig. 3.

This assumption of CSIT is common in hybrid precoding designs for mmWave massive MIMO systems [9, 10]. We can represent the transmitted signal as

$$x = F_{RF} F_{BB} s \qquad (1)$$

Where we express the baseband precoder as

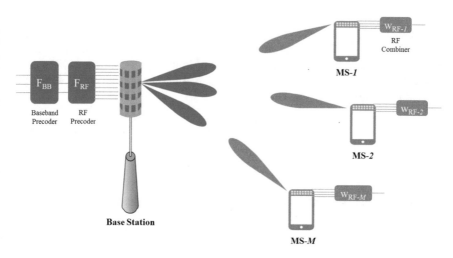

Fig. 3 MS sharing channel information with BS

$$F_{BB} = \left[f_{BB1} f_{BB2}, \ldots f_{BBk} \right] \in \mathbb{C}^{K \times K} \tag{2}$$

The data stream vector is considered as $s \in C^{K \times 1}$ and for convenience $E\left\{ ss^H \right\} = \dfrac{P}{K}.I$, where P is the total transmit power and we assume each data stream is allocated with equal power. The analog precoder $F_{RF}s \in \mathbb{C}^{N_t \times K}$ is realized using phase shifters and hence each element of F_{RF} is based on constant modulus constraint. When fully connected hybrid architecture is used, then F_{RF} can be written as

$$F_{RF} = \left[f_1 f_2 \ldots f_K \right] \tag{3}$$

Here $f_k \in \mathbb{C}^{N_t \times 1}$, and it is noted that, F_{RF} is a full matrix with all elements as non-zero and normalized such that

$$\left| f_k(m) \right|^2 = \frac{1}{N_t} \tag{4}$$

The baseband precoder is normalized to fulfill the total power constraint as $\left\| F_{RF} F_{BBF} \right\|^2 = K$. At the receive terminal, each UE k is having N_r^{RF} RF chains and therefore its combiner $W_k^H \in \mathbb{C}^{K \times N_r^k}$ satisfying (5).

$$\left| W_k(m) \right|^2 = \frac{K}{N_r^k} \tag{5}$$

The received signal at the user equipment can be written as

$$r_k = W_k^H H_k F_{RF} f_{BBk} s_k + W_k^H H_k F_{RF} \sum_{j \neq k} f_{BBj} s_j + W_k^H n_k \tag{6}$$

where $H_k \in \mathbb{C}^{N_r^k \times N_t}$ represents the mmWave channel between BS and UE k and n_k is the additive noise which follow circular symmetric distribution with zero mean and σ^2. I as variance. Therefore by the principle of unitary transformation, $W_k^H n_k$ also has the same distribution as that of n_k. The signal to interference plus noise of the received signal for k^{th} UE can be derived as

$$\gamma_k = \frac{\dfrac{P}{K} \left\| W_k^H H_k F_{RF} f_{BBkF} \right\|^2}{\displaystyle\sum_{j \neq k} \dfrac{P}{K} \left\| W_k^H H_k F_{RF} f_{BBjF} \right\|^2 + \sigma^2} \tag{7}$$

The achievable sum rate can be expressed while input is assumed to be Gaussian

$$R = \sum_{k=1}^{K} \log_2 \left(1 + \gamma_k\right) \tag{8}$$

1.2   mmWave MIMO Channel Model

The mmWave channels render limited scattering [11, 12]. Considering this characteristics, the channel model include L_u independent multipaths for each H_k [13, 14], where $L_u < N_t$ for limited scattering. Then the channel between BS and each UE can be represented as

$$H_k = \sqrt{\frac{N_t N_r^k}{L_u}} \sum_{l=1}^{L_u} \alpha_{k,l} \cdot a_{UE}\left(\theta_{k,l}^{UE}, \varnothing_{k,l}^{UE}\right) a_{BS}^H \left(\theta_{k,l}^{BS}, \varnothing_{k,l}^{BS}\right) \tag{9}$$

where $\alpha_{k,l}$ is the complex path coefficient and is considered to follow the standard i.i.d complex Gaussian distribution that produce the Rayleigh component. $a_{UE}\left(\theta_{k,l}^{UE}, \varnothing_{k,l}^{UE}\right)$ and $a_{BS}^H\left(\theta_{k,l}^{BS}, \varnothing_{k,l}^{BS}\right)$ are array response vectors at receiver and transmitter respectively, where $\theta_{k,l}^{UE}\left(\varnothing_{k,l}^{UE}\right)$ and $\theta_{k,l}^{BS}\left(\varnothing_{k,l}^{BS}\right)$ azimuth (elevation) angles of arrival and departure (AoAs/AoDs) of lth path. For uniform linear arrays (ULAs), only the azimuth AOAs and AoDs are considered and the array response vector is given by

$$a_{ULA}\left(\theta\right) = \frac{1}{\sqrt{N}} \left[1, e^{jkd\sin(\theta)}, \ldots, e^{j(N-1)kd\sin(\theta)}\right]^T \tag{10}$$

where $k = \dfrac{2\pi}{\lambda}$ with λ as wavelength of the carrier signal, and d is the spacing between the antenna elements. For uniform planar array antennas, the array response vector is given by

$$a_{UPA}\left(\theta,\phi\right) = \frac{1}{\sqrt{N}}\left[1, e^{jkd\left(m\sin(\phi)\sin(\theta)+n\cos(\theta)\right)}, \ldots, e^{jkd\left((W-1)\sin(\varnothing)\sin(\theta)+(H-1)\cos(\theta)\right)}\right]^{T}$$

where $0 \leq m \leq W - 1$, $0 \leq n \leq H - 1$, and $N = WH$. $N = N_t$ for a_{BS} and $N = N_r^k$ for a_{UE} of H_k.

1.3 Problem Formulation

Enlarging the system data rate to the maximum capacity in a multi-user mmWave downlink system is a critical task. The central challenge arises in merging the analog and digital precoders, which demands unique and distinguished constraints to be satisfied while realizing a precoder. The implementation of the precoders as described in (6) needs a complex matrix computed to obtain the product of baseband and RF precoding matrices. Further, the character of constant modulus constraint concerning analog phase shifters demands the entries in the RF precoding matrix to possess equal norms so that its columns can be extracted as the RF beamforming vectors. Furthermore, as described, determining the precoding matrices is comparable to answering an optimization objective with non-convex possible constraints, which lacks in providing a generic solution [15]. Therefore, only estimated solutions can be provided for the fundamental optimization issue, and hence, the sub-optimal but less-complex hybrid precoding/combining designs are appreciated. In this work, we assume that the mmWave link is sparse in the angular domain, i.e., there may be only a few of AoAs of AoDs.

The traditional MIMO channel can be decomposed as $H = U \sum V^H$, where the matrices U and V are semi-unitary such that $U^H U = I$ and $V^H V = I$ and $\sum$ is the diagonal matrix of non-negative values [16]. For the classical MIMO, we can set the baseband precoder equal to V and RF precoder to I or they can even be interchanged [17]. But for mmWave massive MIMO, the ideal baseband/RF precoder must be set to $A_t V$. The transmit array response vectors in transmit array matrix A_t form a basis for the column space of H^H [18]. The matrix A_t need to be estimated either by constructing a dictionary matrix or by applying the principle of scattering. Therefore, the best precoder approximation problem can be formulated as $\arg\max_{F_{BB}} \left\| V - A_t \widetilde{F_{BB}} \right\|_F^2$.

At the receiver, the received vector Y is acted upon by combining weights. The combiner weights need to be computed so as to reduce the minimum mean square error (MMSE) of the receiver. The HBF system aims to deliver greater spectral efficiency, the precoding and combining weights decides the performance of the HBF system. Therefore, the optimization problem is formulated so to optimally compute these weights. As the RF beamforming and combining coefficients are used only to control the direction of beams, there are few additional constraints in the optimization problem to compute these weights. Theoretically, the combination of $F_{BB} \times F_{RF}$ and $W_{BB} \times W_{RF}$ that are produced by the HBF method are expected to be near estimates of F and W which are computed without any constraints.

With this considerations, the hybrid precoding/combining design problem can be expressed as a sparse approximation problem, and a modification of the matching pursuit algorithm named extended orthogonal matching pursuit is developed to efficiently produce the RF/baseband precoding and combining matrices. This work confirmed that hybrid precoding with reduced hardware and cost can considerably rescue performance accretions pertinent to digital baseband methods. Thus hybrid precoding marks itself as an encouraging solution for mmWave systems. The recommended hybrid precoder/combiner produces spectral efficiencies that are very near to those delivered by the optimal unconstrained digital precoding solution (Fig. 4).

2 Extended Orthogonal Matching Pursuit Method

In this section, we discuss an effective mechanism that considers the beam grids for computing the array response matrices. Subsequently, we propose an extended simultaneous orthogonal matching pursuit algorithm (ESOMP). The transmit array response matrix can be constructed by considering the angular grid $\emptyset_T$ of size G, with $\theta \in \emptyset_T$, $1 \leq i \leq G$, and $G \geq N_t$. The dictionary matrix A is constructed as in (3).

$$A = \left[a(\theta_1) a(\theta_2) a(\theta_3) \ldots a(\theta_G) \right] \text{ and}$$

$$a(\theta_1) = \begin{bmatrix} 1 \\ e^{-j\frac{2\pi}{\lambda} d \cos\theta_l} \\ \vdots \\ e^{-j\frac{2\pi}{\lambda}(N_t - 1) d \cos\theta_l} \end{bmatrix} \tag{12}$$

Fig. 4 Angular grids   **Grid of Beams**

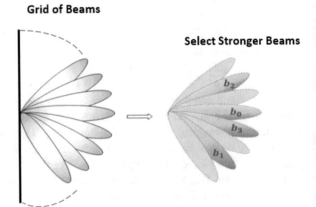

Figure 6 represents the formation of angular beam grids across the entire region of coverage. The transmitter system selects only the specified number of directions in which beams are stronger and receivers are located. The transmission occurs only on those selected directions. This selection of stronger beams happens accurately if the precoder is designed with the best approximation. The precoder optimization problem is formulated by estimating a block sparse matrix. The optimal baseband precoder is determined by extracting the non-zero rows from the block sparse matrix. The RF precoder is formed from the columns of the array response matrix which corresponds to the non-zero rows of the block spark matrix. The RF combiner is determined from the receive array response matrix which is computed using the least-squares solution. The baseband combiner is formed by processing the RF combiner along with the channel covariance matrix.

The optimization of precoder (baseband and RF precoding) and combiner (RF combining and baseband combining) can be achieved through the proposed scheme as illustrated in the algorithm. The proposed scheme is employed for a uniform linear and planar array antenna systems, and the performance is presented for both the antenna systems. The flow of research method is as depicted in the flowchart of Fig. 5.

Extended SOMP Algorithm: Precoder Optimization

$\mathbf{F}_{RF}^{(0)} = [\ ]$,

$[\mathbf{U}, \mathbf{S}, \mathbf{V}] = \mathbf{svd}(\mathbf{H}), \mathbf{F}_{opt} = \mathbf{V}(:, 1 : N_s)$

$\mathbf{F}_{res}^{(0)} = \mathbf{F}_{opt}$

for $1 \leq k \leq N_t^{RF}$

$\psi = \mathbf{A}_t^H \mathbf{I}_{N_t} \mathbf{F}_{res}^{(k-1)}$

$i(k) = \text{agr max } [\psi\psi^H]_{l, l}$

$\mathbf{F}_{RF}^{(k)} = \mathbf{A}_t\left[:, i(k)\right]$

$\mathbf{F}_{BB}^{(k)} = \left(\left(\mathbf{F}_{RF}^{(k)}\right)^H \mathbf{F}_{RF}^{(k)}\right)^{-1} \left(\mathbf{F}_{RF}^{(k)}\right)^H \mathbf{F}_{opt}$

$\mathbf{F}_{res}^{(k)} = \dfrac{\mathbf{F}_{opt} - \mathbf{F}_{RF}^{(k)}\mathbf{F}_{BB}^{(k)}}{\left\|\mathbf{F}_{opt} - \mathbf{F}_{RF}^{(k)}\mathbf{F}_{BB}^{(k)}\right\|_F}$

end

$\mathbf{F}_{BB} = \dfrac{\sqrt{N_s}\mathbf{F}_{BB}}{\left\|\mathbf{F}_{RF}\mathbf{F}_{BBF}\right\|}$

$\mathbf{F}_{BB} = \mathbf{F}_{BB}^T, \mathbf{F}_{RF} = \mathbf{F}_{RF}^T$

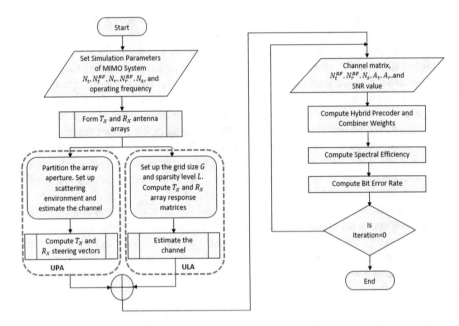

Fig. 5 Flowchart of research method

Extended SOMP Algorithm: Combiner Optimization

$$\mathbf{W}_{RF}^{(0)} = [\ ],$$

$$\mathbf{W}_{res}^{(0)} = \mathbf{W}_{MMSE}$$

$$\mathbf{R}_{SS} = \mathbf{I}_{N_s}, \mathbf{R}_{YY} = \mathbf{H}\mathbf{F}_{RF}\mathbf{F}_{BB}\mathbf{R}_{SS}\mathbf{F}_{BB}^{T}\mathbf{H}^{T} + SNR^{2}\mathbf{I}_{N_r}$$

for $1 \leq k \leq N_r^{RF}$

$$\psi = \mathbf{A}_r^{H}\mathbf{R}_{YY}\mathbf{W}_{res}^{(k-1)}$$

$$i(k) = agr\ max\ [\psi\psi^{H}]_{1,1}$$

$$\mathbf{W}_{RF}^{(k)} = \mathbf{A}_r\left[:,i(k)\right]$$

$$\mathbf{W}_{BB}^{(k)} = \left(\left(\mathbf{W}_{RF}^{(k)}\right)^{H}\mathbf{R}_{YY}\mathbf{W}_{RF}^{(k)}\right)^{-1}\left(\mathbf{W}_{RF}^{(k)}\right)^{H}\mathbf{R}_{YY}\mathbf{W}_{MMSE}$$

$$\mathbf{W}_{res}^{(k)} = \frac{\mathbf{W}_{MMSE} - \mathbf{W}_{RF}^{(k)}\mathbf{W}_{BB}^{(k)}}{\mathbf{W}_{MMSE} - \mathbf{W}_{RF}^{(k)}\mathbf{W}_{BB}^{(k)}\ _{F}}$$

end

$$\mathbf{W}_{BB} = \mathbf{W}_{BB}^{*}, \mathbf{W}_{RF} = \mathbf{W}_{RF}^{*}$$

3   Results and Discussions

This section presents the simulation results of the proposed scheme under an $Nr \times Nt = 16 \times 64$ mmWave MIMO system. The two types of antenna orientation are considered, namely, uniform planar array (UPA, 8×8 at transmitter and 4×4 at the receiver) and uniform linear array (ULA) with antenna spacing of half the wavelength. The transmitter and receiver are assumed to a maximum of 6 RF chains, and the simulations have been shown for 4 and 6 RF chains. The results are based on the channel model described in [19], which is assumed 6 scattering clusters that are distributed randomly in a scattering environment. Each cluster consists of 8 scatters that are closely located with 5° of angle spread for all 48 scatters. The path gain for each scatterer is obtained from a complex circular symmetric Gaussian distribution with uniformly distributed mean angles and angular spread of 5°. The popular SOMP algorithm [20] for the hybrid beamforming scheme is compared with the proposed hybrid beamforming (ESOMP method) scheme, and the digital beamforming method is included for ease of comparison. All simulation results are the average of more than 100 independent channel realizations.

Figures 6 and 7 represent the antenna geometry of UPA at transmitter and receiver. Spacing between the adjacent antenna elements is $\frac{\lambda}{2}$ uniformly. The 2D antenna array is positioned in X-Y plane and the transmission and reception happens in broadside, i.e., in Z direction. The transmitter radiation response of UPA antenna system is exhibited with respect to analog (Fig. 8) and hybrid beamforming (Fig. 9) techniques.

The analog beamforming produces the radiations in only a single direction which is most dominant. As an ABF can only serve a single user or only a single data

Fig. 6 UPA geometry at BS

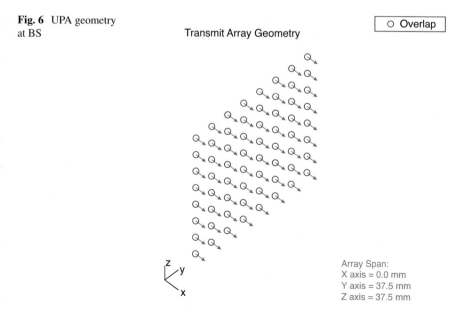

Transmit Array Geometry

○ Overlap

Array Span:
X axis = 0.0 mm
Y axis = 37.5 mm
Z axis = 37.5 mm

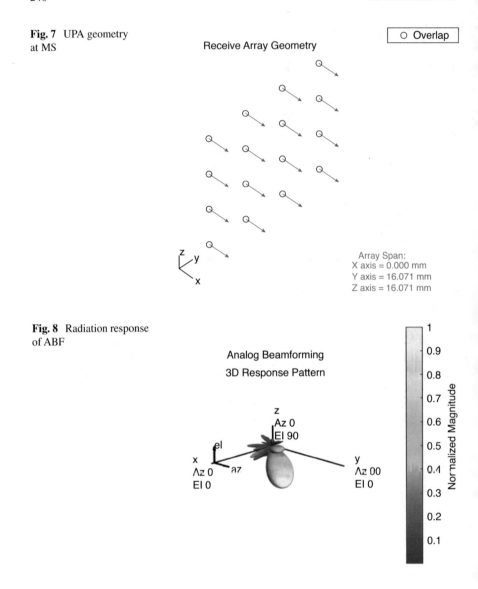

Fig. 7 UPA geometry at MS

Fig. 8 Radiation response of ABF

stream at a given instant, hybrid beamforming produces the multiple beams serving multiple users/multiple data streams. Figures 10 and 11 illustrate the spectral efficiency against the signal-to-noise ratio (SNR) for different algorithms with ULA and UPA antennas, respectively. The proposed HBF scheme performs significantly better than the SOMP algorithm, and the achieved spectral efficiency is close to the DBF. Finally, we examine the BER performance, presented in Figs. 12 and 13. The BER of the proposed scheme ensures that the ESOMP algorithm is more robust as compared to the existing SOMP method. The simulation results reveal that the proposed HBF algorithm outperforms the existing SOMP method.

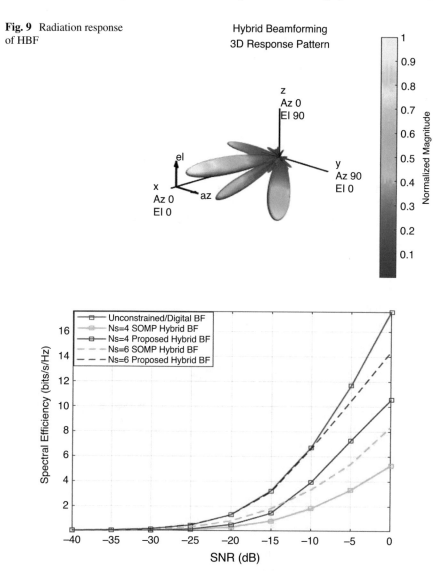

Fig. 9 Radiation response of HBF

Fig. 10 Spectral efficiency for ULA

4 Conclusion

This paper presents a hybrid beamforming algorithm that produces the optimum precoding and combining weights for mmWave MIMO communications for the downlink scenario. While reducing the hardware complexity, cost, and power consumption, the proposed scheme achieves the improved spectral efficiency and bit error rate performance compared to the SOMP algorithm. Therefore, the scheme

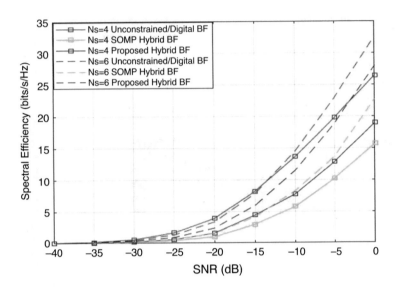

Fig. 11 Spectral efficiency for UPA

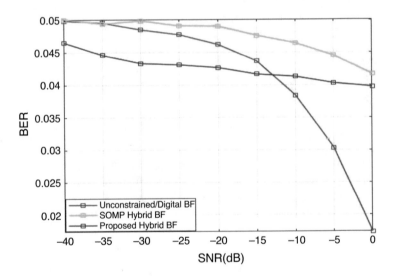

Fig. 12 BER Performance of ULA

proposed in this paper is beneficial and can be employed in the implementation of the multi-user mmWave communication system. This work can be extended further by using non-orthogonal multiple access in combination with mmWave MIMO. The mmWave massive MIMO hybrid beamforming systems improve the performance at a reduced cost. However, the proposed scheme does not address serving multiple

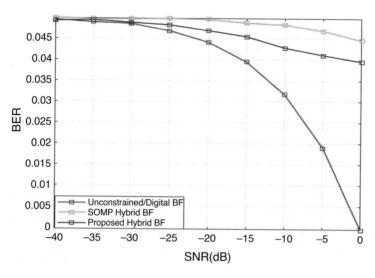

Fig. 13 BER Performance of UPA

users in each beam and thus leads to an inadequate number users being served, as the number of users cannot be more than the number of RF chains at a given instant. To further increase the spectrum efficiency, further study reveals the need of combining NOMA with mmWave massive MIMO systems. Therefore, the so-called mmWave massive MIMO-NOMA systems promise to serve multiple users per beam at the same time and frequency by applying superposition coding within the same beam at the transmitter and successive interference cancellation (SIC) at the receivers.

References

1. M. Cudak et al., Moving towards mmwave-based beyond-4G (B-4G) technology, in *2013 IEEE 77th Vehicular Technology Conference (VTC Spring)*, (IEEE, 2013)
2. E.G. Larsson et al., Massive MIMO for next generation wireless systems. IEEE Commun. Mag. **52**(2), 186–195 (2014)
3. M.H. Elmagzoub, On the MMSE-based multiuser millimeter wave MIMO hybrid precoding design. Int. J. Commun. Syst. **33**(11), e4409 (2020)
4. Y.-Y. Lee, C.-H. Wang, Y.-H. Huang, A hybrid RF/baseband precoding processor based on parallel-index-selection matrix-inversion-bypass simultaneous orthogonal matching pursuit for millimeter wave MIMO systems. IEEE Trans. Signal Process. **63**(2), 305–317 (2014)
5. X. Wu, D. Liu, F. Yin, Hybrid beamforming for multi-user massive MIMO systems. IEEE Trans. Commun. **66**(9), 3879–3891 (2018)
6. S. Malkowsky et al., The world's first real-time testbed for massive MIMO: Design, implementation, and validation. IEEE Access **5**, 9073–9088 (2017)
7. S. Blandino et al., Multi-user hybrid MIMO at 60 GHz using 16-antenna transmitters. IEEE Trans. Circuits Syst. Regul. Pap. **66**(2), 848–858 (2018)

8. S. Sun et al., MIMO for millimeter-wave wireless communications: Beamforming, spatial multiplexing, or both? IEEE Commun. Mag. **52**(12), 110–121 (2014)
9. J. Mo, R.W. Heath, Capacity analysis of one-bit quantized MIMO systems with transmitter channel state information. IEEE Trans. Signal Process. **63**(20), 5498–5512 (2015)
10. S.A. Nezamalhosseini, L.R. Chen, Optimal power allocation for MIMO underwater wireless optical communication systems using channel state information at the transmitter. IEEE J. Ocean. Eng. **46**, 319 (2020)
11. R. Rajashekar, L. Hanzo, Iterative matrix decomposition aided block diagonalization for mm-wave multiuser MIMO systems. IEEE Trans. Wirel. Commun. **16**(3), 1372–1384 (2016)
12. C. Huang et al., Iterative channel estimation using LSE and sparse message passing for mmWave MIMO systems. IEEE Trans. Signal Process. **67**(1), 245–259 (2018)
13. X. Wang et al., Millimeter wave communication: A comprehensive survey. IEEE Commun. Surv. Tutorials **20**(3), 1616–1653 (2018)
14. K.V. Mishra et al., Toward millimeter-wave joint radar communications: A signal processing perspective. IEEE Signal Process. Mag. **36**(5), 100–114 (2019)
15. O. El Ayach et al., Spatially sparse precoding in millimeter wave MIMO systems. IEEE Trans. Wirel. Commun. **13**(3), 1499–1513 (2014)
16. J. Lee, G.-T. Gil, Y.H. Lee, Channel estimation via orthogonal matching pursuit for hybrid MIMO systems in millimeter wave communications. IEEE Trans. Commun. **64**(6), 2370–2386 (2016)
17. Z. Wang et al., Iterative hybrid precoder and combiner design for mmWave multiuser MIMO systems. IEEE Commun. Lett. **21**(7), 1581–1584 (2017)
18. S. Adnan et al., Sparse detection with orthogonal matching pursuit in multiuser uplink quadrature spatial modulation MIMO system. IET Commun. **13**(20), 3472–3478 (2019)
19. S. Haghighatshoar, G. Caire, Enhancing the estimation of mm-wave large array channels by exploiting spatio-temporal correlation and sparse scattering, in *WSA 2016; 20th International ITG Workshop on Smart Antennas*, (VDE, Frankfurt am Main, 2016, March), pp. 1–7
20. O. El Ayach, S. Rajagopal, S. Abu-Surra, Z. Pi, R.W. Heath, Spatially sparse precoding in millimeter wave MIMO systems. IEEE Trans. Wirel. Commun. **13**(3), 1499–1513 (2014)

Biometric Identification System: Security and Privacy Concern

Gurinder Singh, Garima Bhardwaj, S. Vikram Singh, and Vikas Garg

1 Introduction

Biometric word comes from the Greek words bio which means (life) and metrikos (measure). We all are aware that all humans have some unique characteristics such as looks, facial parameters, gait, or voice, and they are being used by other humans to identify other [1]. In today's fast changing and technology-dependent world, a lot of applications and systems requires a system for dependable verification to ensure the identity of a person to enable them to access the data and facility or confirm identity [2].

The unique feature of human body that separates them from other human body is being used to confirm their identity from a long time [3]. In technical term, we have been using security passwords and identity cards to confirm genuine access to the protected systems, but they can be ruptured easily hence dependency on them is very low [24]. The biometric characteristics of a living body can't be forgotten or stolen, and forging anyone will be very difficult. The biometric system must be a pattern-identification application that will identify a living human depending on feature or characteristic which any human possesses [4].

That feature or characteristic may be saved in a secured database in the form of template (or saved in the ID card of the individual) after being extracted from a human body [5]. Any biometric system which is dependent on physical characteristics is usually more dependable as compared to the systems which depend on behavior of an individual [6].

Biometric system solves two main purposes:

G. Singh · G. Bhardwaj · S. V. Singh (✉) · V. Garg
Amity University, Noida, Uttar Pradesh, India
e-mail: gsingh@amity.edu; gbhardwaj@gn.amity.edu; svikram@gn.amity.edu;
vgarg@gn.amity.edu

© The Author(s), under exclusive license to Springer Nature Switzerland AG 2021
S. Awasthi et al. (eds.), *Artificial Intelligence for a Sustainable Industry 4.0*,
https://doi.org/10.1007/978-3-030-77070-9_15

1. Verification
2. Identification

- Verification: Verification means to compare the saved templates with respect to the claimed identity.
- Identification: Identification means comparing the users data with the saved biometric information if form of template in the main database.

Thus above two operations are to be dealt separately. This system begins with the selection of a special human characteristic as an input data, and this data is to be used for the verification and identification of any person in a security system [7].

Characteristics of Biometric System: Biometric Systems have some very special characteristics such as:

1. Universality: Every living person possesses biometric traits; hence biometric security system is a universal means it may be applicable to all.
2. Individuality: Individuality is one of the most important properties of biometric system as it will ensure that there is no replica and every different person has their individual biometric traits.
3. Security: Biometric traits are one of the most secured one.
4. Accuracy: Biometric systems are very accurate as biometric traits of a person will never match with some traits of some other person.
5. Reliability: Reliability is one of the measure concerns of any security system. Biometric systems are very reliable, and they enhance the overall security of the system.
6. Acceptability: Biometric systems are acceptable across the globe, gender, and race of human. They are being used in higher as well as lower financial segment of people (Fig. 1).

Some more biometric characteristics have also been marked, and we are using them in different systems to protect and secure our systems.

Fig. 1 Characteristics of biometric system

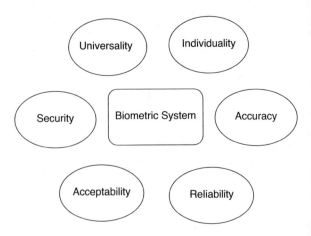

2 Biometric Modalities

Biometric which are being used as most commonly are fingerprints scan, face recognition system, IRIS scan, palm and fingers geometry, and voice. We scan/record the biometric trait or template as the reference and save them in main database which is letter on used for verification of persons by matching the input with saved records [8].

2.1 Fingerprint Scan

The shape of ups and down on the fingers has been used in number of applications from very old time [9]. Two persons can't have similar fingerprint pattern. Nowadays, we are taking scans of many fingers of a person so that multiple prints can have additional information resulting in enhanced security [10] (Fig. 2).

2.2 Human Face

Human face is a unique characteristic that can be utilized in biometric system for identification of a person. Technology for face recognition is popular and is easy to use because it does not require physical connection or touch between the person and security system; also this system does not require any costly hardware [11]. This technology can be used in different security systems (Fig. 3).

Fig. 2 Fingerprint image

Fig. 3 Human face X-ray
image

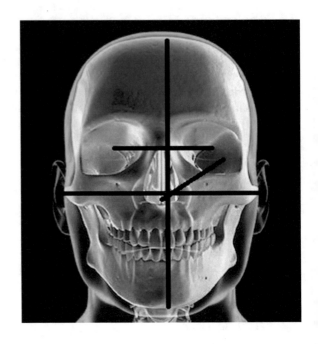

2.3   Hand Measurement

Shape of hands of a human, size of palm, fingers, length, etc. are measured and
saved as a template in this model [12]. This model is very easy to use and simple.
This system is mainly used for verification purpose. This verification system is however too large for laptop, computer, etc. [13] (Fig. 4).

2.4   Palm Scan

Palm scan is a set of fingerprints and hand. Palm contains ups and down lines. This
arrangement is also very large in the size. They are more useful in forensic science
as many scans can be collected from any crime location [14] (Fig. 5).

2.5   IRIS Scan

Iris is present in eyes in circular shape. It is a very thin layer. Iris is the controller of
dimension of pupil. Iris also limits the light which is allowed to go in retina. This
protects retina of the eyes. Iris may be of different color depending on the genes of
a person [16]. The color of eyes of an individual is dependent on the color of Iris

Fig. 4 Hand measurement image

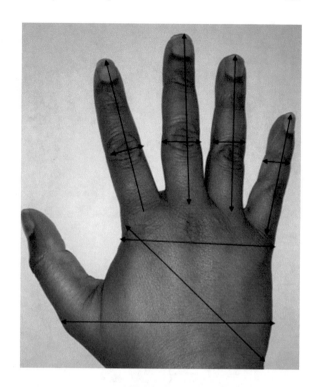

Fig. 5 Palm scan image

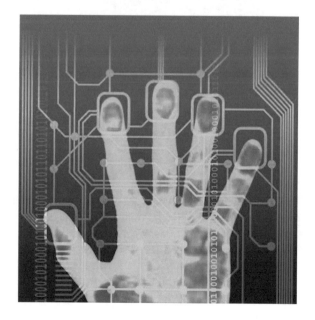

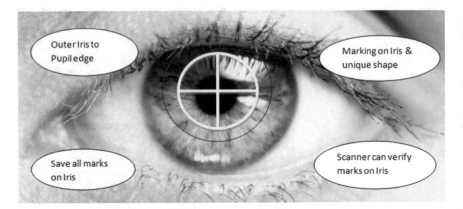

Fig. 6 Retina image

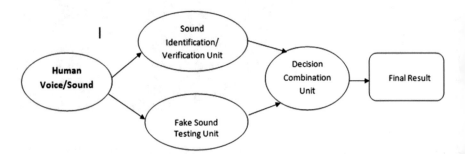

Fig. 7 Voice process

such as green, brown, blue, gray, hazel, violet, etc. The pattern of iris is also different in eye to eye and for individual [17] (Fig. 6).

2.6 Voice

Voice is a combination of behavioral and bio characteristics of a human. The voice sound of any human is dependent on many parts of human body [18] (mouth, nose, lips, vocal cords, and so on). Voice is also dependent on age, mental stage, location, and also medical conditions. We may record the voice of a person in different conditions, and it can be used for verification purpose. We can have special spoofing detection module to avoid any fake voice verification [19] (Fig. 7).

All the above biometric modalities may be used depending on the type of security required and the device. We can use more than one biometric method to protect the systems with higher risk.

3 Use of Software for Biometric Security

Use of software and middleware for biometrics helps us in making the link among instructions and the services. It also gives instruction with the use of multiple processes [20]. It also supports the biometric systems and databases to run effectively across the network. Such software connects biometric devices with the computers/devices, and they work comfortably with each other in network [21]. This software also provides flexibility to combine all the applications placed at the server or main database.

4 Various Issues in Biometric System

The biometric verification and authentication systems are likely secure, although there are two different and unique features with 2 special technologies [22].
Important security issues related to the biometrics are mentioned below.

4.1 The Algorithm (Encryption) Used in Biometrics Is Weak

The stored biometric data is coded using a special algorithm in order to save and secure the data in a main database. All such algorithms possess some specific security standards [23]. Since the calculative power of the computer is enhancing day by day, this will help in finding the solution in very less time and calculations.

4.2 Size of Biometric Database Increases with Time

Since multiple features and samples are recorded for one user, size of database increases with time when recorded for many users [2].

4.3 Size of Storage and Key Size

The key for conversion of encrypted data is a crucial and important unit, which must be kept undisclosed. Hence, this key must be impenetrable [4]. However, with the increase in the size of this key, processing time for computation also increases.

4.4 Cyber/Criminal Attacks

Users can be threatened, kidnapped, or wrongly assessed, or their data can be stolen by cyber-attacks on the databases.

4.5 Factors Related with Environment

In biometric systems (especially in fingerprint), there can be degradation in quality when fingers are shrink or fingertips got wrinkled, it generally happens because of working with fingers in soil, water, or chemicals (Reports TechCrunch web). As in the case of maritime domain.

Hence, the correctness of the biometric systems is very much related with the various environmental factors.

4.6 Foreign Particles

Few foreign particles like dust and moisture on the scanner surface can wrongly present the data and can affect the working of scanner.

4.7 Attack on Comparator

The comparator can also be attacked by any virus that will always create wrong match scores during authentication regardless of the user's data present at the biometric sensor.

4.8 Template Got Attack by Virus

As we create a template for enrollment, this template is kept secured in the main database. The virus attack can change the saved template in the database; also it can replace it with some other template.

4.9 Medium of Communication

Any saved data which is being transported via a communication link is on risk for interception, recording, and modification, and it may again be inserted into the system with some wrong intention.

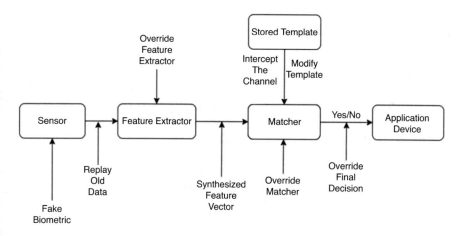

Fig. 8 Possible attacks on any biometric security system

4.10 *Decision May Be Altered*

The final result given by our biometric device can be hacked and altered by a virus or hacking program. This will damage entire security system (Fig. 8).

5 Methods of Dealing with Issues in Biometric System

Using special device (Anti-Spoofing) and life detection test: All biometric systems possess high risk to the spoofing attacks. For any biometric system which is dependent on the fingerprint scanning, we can use a special device which will work by sensing live pulses from human body or fingers. By this, the finger can be verified for liveness. This technique is not very costly and simple and can be used to enhance the security status of the biometric scanners. It also eliminates spoofing attacks from dummy and artificial fingers. The finger is of a live human being can be checked by comparing original and false fingerprints depending on the quality and quantity of the pore. Pore quantity of a genuine fingerprint will be much more in comparison with that on a false fingerprint. Hence, in order to verifying that the fingerprint of a live person, at least three detection pore patterns with the ridges must be collected to determine the sample fingerprint data that is taken from a genuine finger else spoofing finger which can be made by material like clay, rubber, silicone, etc. We can also determine fake fingerprints by quantifying the minutiae count. Ridge endings and differentiation available on a fingerprint is known as Minutiae points.

5.1   Multimodal Biometrics

A multimodal biometrics is a set of two or more biometric modality in any single verification system. This model can be used when the protected data is very sensitive and important. Introducing different modalities results in enhanced recognition accuracy. It also improves the privacy of the system as it is not so easy to forge many biometric traits together. Multimodal biometric systems ask the end user to store different biometric traits randomly who will ensure genuine fingerprint detection to protect data from spoofing or hackers.

Multimodal biometric system is of universal type as sometimes a person is not able to give any specific form of biometrics due to particular disability or illness; the system can collect some other form of biometric for verification (Fig. 9).

5.2   Touchless Fingerprint

As conventional fingerprint recognition system, there are several issues encountered in this method for example wet fingers, dry fingers, dust particles, etc. Such issues can cause changes in the sample image collected. In order to avoid such issues and drawbacks, we may use a high resolution digital camera to capture the fingerprints or images. There are three main stages for this system; they are pre-processing, feature extracting, and comparison.

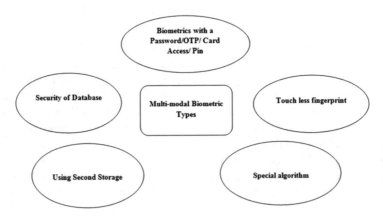

Fig. 9 Multimodal biometric types

5.3 Using Special Algorithm Having Enhanced Security Specification

In order to confirm the security of 112 bits, the **Rivest-Shamir-Adleman (RSA)** algorithm needs key of the size of 2048 bits. As the computing power of the computers is increasing swiftly, a special and secure algorithm created today can be easily ruptured in the coming future. Therefore, the selection and regular update of the designed algorithms are very important.

5.4 Using Second Storage

By using memory chips or hard disc having storage to keep secure the coded biometric data as a second option for the main database. In this method, important data get secured, and the chance of loss of all secured data to ghost virus or hacker is very low.

5.5 Security of Database

The stored biometric data in the provided database must be kept secured and stored so that even if any hacker or external person gets the database, the database should not be accessed by them by disabling them to copy the real pattern (may be biometric) to the system.

5.6 Biometrics with a Password/OTP/Card Access/Pin

Using an extra verification system which will identify and verify individuals depending on card access, passwords, or pin but as it is extremely easy to forge PINs and passwords, it will be a weak security arrangement. But if the biometric system is joined with OTP/PIN/Card/Password to verify/ confirm a user, the system will have higher security. One-time password can also be added with biometric in order to making our biometric system secure.

6   How Addition of Biometrics Enhance the Security

For confirmation and enhancement of security, the most important parameter is confirmation of identity and their consent it is therefore many organizations are using biometric attendance or identification/Verification device. Biometric system is a key element for verification, and it can be used for multifactor verification. It is being used for tracking, attendance of individual, managing time limit, etc. In organizations, we are using biometrics for permitting entry and exit, most commonly fingerprint verification.

The following are the most important benefits with growth of multifunction biometric authentication.

6.1   Fast and Reliable Verification System

Biometric authentication and verification system are very fast and very reliable; passwords, OTP, and security codes are very fast but generic. Any person having these codes may access the security system and breach the security. However, in the case of biometric security system, the password is biological identity; hence the chance of forgery is very less. Scanning iris or face recognition system can be used very easily for enhancing the security system.

6.2   Accountability of the Individual

In biometric security system, we have firm information about the entry, exit, and availability of the individual staff. In case of any unwanted situation, we have documented proof regarding the presence of any individual. This data can be used, analyzed, and presented to any mass very efficiently and easily.

6.3   High Efficiency

Nowadays, in every organization whether it is big or small, they demand a very secure and efficient security system. With the addition of biometric verification, the security of our system is enhanced, and also it makes an easy and efficient security system which also manages many important tasks also such as attendance preparation for salary preparation. It will also enable the employees to not carry their ID cards everywhere.

6.4 Convenient

A major advantage of addition of biometrics in security system is convenience as in this we don't need to reset the password; we don't need to carry a separate device/ ID card. We just need one-time biometric template collection such as fingerprints, iris, and facial recognition, and our manpower can be identified at any step. Thus this addition makes our security system convenient also.

6.5 Adoptability of Future Growth of the Organization

With the growth of the organizations, security parameters and number of employees also grow very rapidly. Biometric verification system provides us the ability to grow. We can save the biometric templates of employees, and verification/identification can be done very easily. No additional machinery or device is required for this. This is a major addition and cost-saving factor also to addition of biometrics.

7 Middleware and Software's Used in Biometrics

A middleware enable us to give a common platform for applications of biometrics and facilitate for development of uses through interconnection of an authentication system with computer. Biometric middleware further provide the feature of interoperability across numerous applications and services that are dependent on the above biometric devices. These middleware also enable us connect instructions and applications via various processes.

Verification and identification by biometrics is not only dependent on the sensors which are used to capture the biometric or biological data of an individual, but it also needs a specially designed algorithms, middleware, and software to store, code, match, and compare the templates, i.e., fingerprint, eye scan, facial scan, etc.

Biometric software or middleware helps in completing the above process which is undoubtfully not an easy task to perform. Protection of biometric data is also a very important task to perform, and this data is always on high risk. To protect this data from losses/hack, this data is coded and kept as a secure database for which algorithms and software are very essential.

The biometric software further enable us to connect biometric machines/scanners with the computer and data network (LAN or WAN) so that they can be accessed from any point, and further they can be operated mutually. With the help of middleware/software biometric, also it allows various application-based software on the separate operating system to become compatible with each other working comfortably.

While performing verification, we math biometric data against a very large size database; these middleware/software help us to find accurate results efficiently in a faction of seconds, as the algorithms and software make it very fast.

Industries having high demand of biometric middleware are as follows.

7.1 FSI

In financial services, biometric systems are used to control and ensure the correct flow of money. We may include biometric in payment system, point of sale verification, authorization for transaction processing, etc.

7.2 Healthcare

In healthcare industries, biometrics may be used for authorization of doctor and staff, monitoring of patients, and maintenance of database/records.

7.3 Manufacturing Industry

With the growth of manufacturing sector, the employee and manpower base of this area increases many folds. Biometrics will help in identification of correct person, and also it will give authority to access the system to authorized persons only.

7.4 Service Industry

Being a vast sector and having highest number of employees biometric security system is becoming an essential for service industry. It helps at both the ends, i.e., employees and customer.

7.5 Government Authorities

In most of the countries, authorities are making biometric database to have a check/control on their citizens. Identification and verification both can be done at any point by government bodies of their citizens for genuine assess.

In addition to above sectors, biometric security systems are being used in most of the sector such as residential, institutions, small and medium industries, etc.

8 Future Biometric Technologies

With the fast-changing world, there are few emerging technologies such as on-spot DNA test and sensing of brain wave. In the next few years, above practices may emerge as a most secure, reliable, and dependable alternative of present available biometric verification system. Many security agencies have already started these methods for verification purpose on borders, on immigration check, or on highly secure area.

8.1 On-Spot DNA Test

On-spot DNA testing is also known as rapid testing or instant DNA testing. This biometric technology can be used for identification as well as for verification. Normal DNA testing is very accurate, but it is a time taking process as it requires to be done at special laboratories and is being used for identification and verification by government authorities after prior approval of statutory bodies.

However, many organizations and scientists of Northwestern University claimed that a Rapid DNA technique has been derived and a handheld device will be available soon for automatic extraction of DNA sample, amplification of data, separation, and detection with human interpretation with technological review of the result. These handheld devices will make it handy and fast. This technology revolution will enable us to identify/verify an individual immediately with highest accuracy.

8.2 Brain Wave Scanning

The root cause of all our emotion, feeling, thought, and action is the network and communication between neurons inside our brains. These brain waves are generated due to electric pulses between communicating neurons. Scanning of brain wave is a new and reliable technology.

Trained operator use sensor by placing it on scalps of a human for sensing the brain waves. The brain waves are then divided into categories (four categories) dependent on their frequency range (given in the Table 1 below).

With the help of the above table by sensing the frequency range of brain wave being produced by brain, we may ascertain the status of mind of a person.

Wave with high frequency ranging from 13 to 30 Hz, i.e., Beta waves are produced when the brain is working or answering of a meaningful question such as driving carefully, appearing for exam, etc. Alpha waves are generated when a person is in some easy stage such as a person resting after hectic schedule, walking in park, fishing, etc. Theta waves are generated when a person is in a relaxing position

Table 1 Brain wave scanning

Brain wave category	Frequency	Condition of mind
Beta (β)	13 Hz & above	Strongly busy
Alpha (α)	8–13 Hz	Resting & cool
Theta (θ)	4–8 Hz	Half sleep
Delta (Δ)	0.5–4 Hz	Deep sleep

from a while. Delta waves are generated from mind when a person is in deep sleeping stage. Further, the frequency from brain wave of a living person could never be zero as this condition will be of a brain dead.

With the help of brain wave on an individual, a brain map is designed, and we can use this as a biometric template. This can be used for verification as well as identification purpose.

9 Advantages of Biometric

Biometric technology is more popular and increasing its importance around the globe. Biometric system is openly accepted by various public sector, private sectors, and multinational organization. It is growing like national identity as its wide use in different sectors include banks, workforce, and identification proofs. Survey shows that people believe and trust more on modern biometric systems in comparison to traditional security systems. There are many reasons behind this revolution of biometric technologies as it is having many advantages. Let's find out certain top benefits of using biometric system.

9.1 Safety and Security

It is very easy to hack the passwords which we used with numbers, alphabets symbols, etc. There are thousands of incidents which are happening every day, and people are losing money constantly. On the other side, biometric technology provides enormous solutions which are unattainable to hack unlike traditional passwords. This system makes life secure for the people and easy with the fighting with security issues.

9.2 Accuracy

Traditional security systems charges regular costing from us in the form of amount of money, time, and resources. Passwords, PIN, and smarts cards are not always accurate, whereas biometric systems work with physical traits of person such as fingerprints, retina, and face recognition that always provide accurate information anywhere, anytime.

9.3 ROI

Biometric technology provides the best ROI as compared to other security systems. This system helps to keep track of thousands of employees of a big enterprise with just one biometric software. On the other side, organization needs to do the same job with various resources and costing becomes much more, so biometric system gives good return on investment.

9.4 Scalability

Biometric systems are extremely scalable for all different types of plans, jobs, and projects. These days biometric system is using by various government projects, employee management, banking security systems, etc. It is feasible and possible because of the scalability of system solution. It gives the enhancement of manage team and work together.

9.5 Screening

As a developed procedure, most visitors travel on visas has fingerprints of two different fingers to scan by inkless device and also digital photograph. These all information helps and assists to the board instructor and gives them data to check whether to accept or reject the traveler. So it helps to screen and take few seconds for the overall processing and makes life easier.

10 Disadvantages of Biometric

10.1 Recognition of Physical Traits

Biometric system works with physical traits like fingerprints, palm scan, etc. We all have the same traits which are unchangeable and can't reset which we can reset in a traditional password system and if sometimes because of some technical reason or changes happen due to some incidents so it would make it difficult to smooth operation.

10.2 Rate of Error

Every system has some loopholes; same with biometric system which also has some rate and types of error first one is false rejection rate which means that system is accepting unauthorized person and false acceptance rate which belongs to the error where system rejects the authorized person. It happens due to physical condition, age, weather, and other related issues.

10.3 Cost

The costs of biometric system are comparatively higher than other traditional security system. It consists the cost of softwares, devices servers, database, etc. with other related machines and equipments which combined huge amount of money.

10.4 Delay

Some biometric systems take more time than expected; the system takes time as it is a long queue of workers who wait to enroll their details in large companies. Employees get hard time for the scanning the biometric system every day. It is hard for people to go through a biometric system before entering in office, school, or other places.

10.5 Complexity

One of the major disadvantages of biometric is extremely technical and complex system that completes the whole process. A person who is non-technical and not tech-friendly can be flopping in the pool of technical people to understand the system.

11 Conclusion

Biometric provides an automatic acknowledgement and recognition of a person based of physical and behavioral traits. These days 80% organizations are tracking their employees presence with biometric where banking industry relies on biometric for customer transaction with their valid physical proof and assure with biometric system. In some cases, fingerprint and face recognition-based system have been worked very effective to protecting information and resources. At present the amount of applications using biometric is limited because of its cost benefit ratio which is not yet have clarity. The future belongs to multidimensional modal biometric technology. Multidimensional modal integrates information at different levels and the most popular at matching score level at one fusion way. Besides with improvement in matching performance, it reflects some problem of spoofing and no universality. It gives new insight of future studies to find some new applications of biometric.

References

1. S. Bevan, S. Hayday, *Attendance Management: A Review of Good Practice*, Report 353 (Institute for Employment Studies, 1998)
2. K. Bredin, Human resource management in project-based organizations. 2008 – Challenges, changes, and capabilities, Doctoral thesis, Department of Management and Engineering, Linköping University, Linköping, 2008
3. C.S. Chikkerur, Online fingerprint verification system, An MTech thesis, Department of Electrical Engineering, and Faculty of the Graduate School of the State University of New York at Buffalo, 2005
4. C. Velazco, Android still most popular smartphone OS, iOS holds steady in second place (on-line), (TechCrunch web, 2011), http://techcrunch.com/2011/11/03/android-still-most-popular-os-ios-holds-steady-in-second-place/. 13 January 2013
5. S. Capoor, Biometrics as a convenience security, Business story 1 December 2006, 48, 50. Retrieved 16 Oct 2012
6. D. Lauren, C. Shane, Enable a fingerprint scanner for android on the Motorola ATRIX 4G (on-line), (developer.com web, 2011), http://www.developer.com/ws/android/devices/enable-a-fingerprint-scanner-for-android-on-the-motorola-atrix-4g.html. 13 January 2013
7. D. Robert, *Human Resource Management* (Heinemann Educational Publisher, 2000)
8. C. Dubin, Biometrics: Hands down, ID management. Security, 1 February 2011, 52, 54
9. J.L. Dugelay et al., Recent advantages in biometric person authentication, in *ICASSP International Conference on Acoustics, Speech and Signal Processing*, (Orlando, Florida, USA, 2002)
10. A.J. Harris, D.C. Yen, Biometric authentication: Assuring access to information. Inf. Manag. Comput. Secur. **10**(1), 12–19 (2002)
11. L. Jiexun, A. Wang, H. Chen, Identity matching using personal and social identity features. Inf. Syst. Front. **13**, 101–113 (2011)
12. S. Kreimer, Matching the right patient to the right record. Hospitals & Health Networks, November 1, 12 (2010). Retrieved 17 Dec 2011
13. D.A. McKeehan, *Attendance Management Program* (The City of Pleasanton, Human Resources, 2002)
14. B.M. Mehtre, Fingerprint image analysis for automatic identification. Mach. Vis. Appl. **6**(2), 124–139 (1993)
15. NSTCS, Biometrics technologies (National Science and Technology Council Subcommittee on Biometrics, 2006). Retrieved 2 July 2012
16. G.C. Ononiwu, G.N. Okorafor, Radio frequency identification (RFID) based attendance system with automatic door unit. Acad. Res. Int. **2**(2) (March 2012)
17. J.K. Ravi, B. Raja, K.R. Venugopal, Fingerprint recognition using minutia score matching. Int. J. Eng. Sci. Technol. **1**(2), 35–42 (2009)

18. F. Sandi Rodiyansyah, rsitektur Sistem Operasi Android (online), (ducnologyweb, 2011), http://educnology.web.id/opensource/rodiyansyah/arsitektur-sistem-operasi-android.   14 Januari 2013

19. V. Shehu, A. Dika, Using real time computer vision algorithms in automatic attendance management systems, in *Proceedings of the ITI 2010 32nd International Conference on Information Technology Interfaces, June 21–24, 2010*, (Caveat, Croatia, 2011)

20. O. Shoewu, O.M. Olaniyi, A. Lawson, Embedded computer-based lecture attendance management system. Afr. J. Comput. ICT **4**(3), 27–36 (2011)

21. L. Tyler, AuthenTec releases SDK for fingerprint sensor applications on Android phones (online), (Ubergizmo web, 2011), http://www.ubergizmo.com/2011/08/authentec-releases-sdk-for-fingerprint-sensor-applications-on-android-phones/. 13 Januari 2013

22. Woodward, J.D., et al., Army Biometric Applications: Identifying and Addressing Sociocultural Concerns. 2001

23. D. Kresimir, M. Grgic, A survey of biometric recognition methods, in *International Symposium Electronics in Marine, ELMAR-2004*, (2004)

24. Z. Rui, Y. Zheng, A survey on biometric authentication: Toward secure and privacy-preserving identification. IEEE Access **99**, 1 (2019)

Enabling Technologies: A Transforming Action on Healthcare with IoT a Possible Revolutionizing

Prasanth Johri, M. Arvindhan, and A. Daniel

1 Introduction

Early smart home developments also used automatic switches, allowing you to transform something or switch something off without raising a finger. However, they did not further integrate with other controls, and their versatility was reduced. This would continue to improve in 1983 as the ARPANET began using the Internet protocol suite (TCP/IP). The protocol standards defined data transmission, routing, and reception standards. This paper set the foundations for the new Internet.

The first wired device to use this new protocol on the Internet was a toaster. John Romkey, a Silicon Valley software developer, designed a prototype computer stand for the 1990 Interop show floor. Romkey placed some slices of bread into the toaster, which he later used to toast his machine. It would be over two decades before anybody used the expression "Internet of Things," but Romkey's magic little toaster revealed the potential for a world of internet-connected things. (Of course, a person always had to put the bread into the oven.) It was a limited release, a beta test; indeed, a harbinger of what was to come [1–3].

The concept "Internet of Things" originally came up in a PowerPoint presentation given in 1999 by Kevin Ashton at Procter & Gamble. Ashton identified a device where sensors operated as the eyes and ears of a computer – an utterly different way for machines to perceive, hears, touch, and view their surroundings.

When the prevalence of home internet increased and Wi-Fi became widespread, the idea of a smart home came to look more realistic. As businesses began to offer an increasing number of smart goods, including "smart" coffee makers, ovens that bake cookies precisely, and refrigerators that replenished old milk, the first of these, the LG refrigerator, was released in 2000. It has detailed information about things

P. Johri (✉) · M. Arvindhan · A. Daniel
Galgotias University, Greater Noida, Uttar Pradesh, India
e-mail: m.arvindhan@galgotiasuniveristy.edu.in; a.daniel@galgotiasuniversity.edu.in

in the store, and it has an MP3 player. This team has spent $20,000. With the decreasing cost of sensor technology, internet-connected devices became increasingly accessible to more users. And the availability of smart plugs, including those manufactured by Belkin, ensured that everyday things could be turned on and off using people's smartphones.

The IoT subsystems consist of limited subsystem modules. First, there's the sensing unit. These sensors may be anything used to capture some kind of data, such as a camera in a smart refrigerator. In certain situations, devices are grouped to collect several data points: a Nest thermostat features a thermometer, but also a motion sensor; it can change the temperature of a room when it detects that nobody's in it (Fig. 1). To understand the data collected by this computer, there is some kind of network connection, and there is a processor to interpret the data. From that info, actions can be taken like buying more milk when the carton in the electronic refrigerator runs out or automatically changing the temperature provided a set of laws [4–6].

1.1 The IoT Is Growing Everywhere

Home appliance automatization: Digital marketer Lauren Fisher refers to the Nest Learning Thermostat, which takes data about the home climate and inhabitants' temperature habits and programs itself to function efficiently. This technological architecture supports the electrical grid, making power more effective.

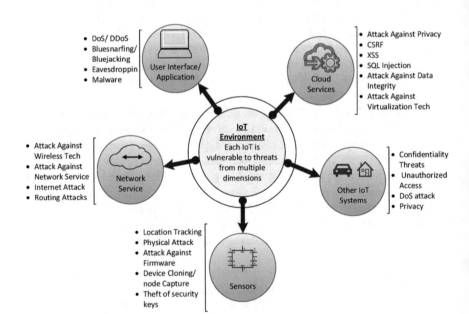

Fig. 1 IoT environment system in multiple dimensions

Alex Brisbourne outlines how the car industry is designing digital applications into cars to include maintenance tracking, fuel and mileage control, driver protection, and other features with little investments but with substantial earning potential. Through adding a cloud-based server to process and make automatic decisions based upon the data, it would help transfer the data to the Internet of Things.

"Technology writer Martyn Casserly cites the London iBus device, which "works with information from over 8,000 buses which are fitted with GPS capabilities besides several other sensors which relay data about the vehicle's position and progress." Bus signposts will announce the time a bus will arrive [7–9].

Just a small percentage of people had smart devices when voice commands were first becoming common. In 2014, Amazon launched the Echo, a voice-activated speaker that has a supportive personal assistant. Four years before, Apple launched Siri, its voice assistant, but Siri lived on your phone, while Alexa lived within the speaker and could monitor any of the "smart" gadgets in your home. By placing a voice assistant as the centerpiece of the smart house, the multifold results become readily apparent: It increased the likelihood that users would purchase more internet-enabled devices and inspired developers to produce more commands for voice assistants to understand.

Much the same year as Amazon launched Alexa and later introduced Homekit, Apple came out with Homekit, a framework built to enable connections between Apple-made smart devices. These unifying voices have moved the world away from a singular purpose-specific automaton and toward developing a more holistic structure of linked objects. Tell Google Assistant to "Goodnight" and the command will set your lights to dim, secure your home, set the alarm, and switch the alarm clock on. LG's SmartThinQ platform allows you to pick cookie recipes from the screen of your smart fridge and set the oven accordingly. The Internet of Things is a handy brand for selling more IoT products. If you still have an Amazon Echo, it might be as well to have the Amazon Echo monitor things [10].

Over the next few years, the number of users online will equal the number of people in the country. David Evans, the former chief futurist at Cisco, reported that "every second, on average, 127 new items are online." Today, the number of devices connecting to the Internet is more than 20 billion, according to figures from Gartner (Fig. 2). The hype and anxiety about the rapidly advancing internet-connected society has reached a fever pitch. With all of these gadgets so advanced, it's easier to monitor the world around you: you can let the delivery guy in the front door or adjust the temperature inside the home, all with a few taps on mobile. It has given us, and the corporations that manufacture these things, more power over us through these objects [11].

The Internet of Things adds the potential of the Internet to devices like light bulbs, thermostats, and doorbells, but it still brings the issues seen on the Internet. Now that consumers have a wide variety of electronic devices able to be wired to a Wi-Fi network, virtually any electronic device in a home can be hacked, or made uselessly. Note the vagaries of connectivity: When your Wi-Fi goes down your computers will no longer work. Network challenges. The smart thermostat can't be turned on with the heat on, because it doesn't work with the smart door lock. Things

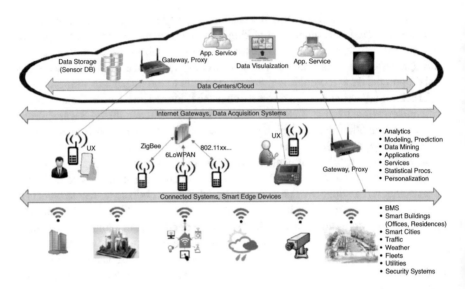

Fig. 2 Real-time remote monitoring system

once simple, now complex, because they are controlled by an Alexa or a smartphone rather than a physical controller. Many of these machines often operate on proprietary software, which means that they will cease to function if the vendor continues providing software updates, ceases to produce the product, or ceases to support the device [12].

Although the risk of bricking an electronic device may be low, linking the device to the internet often raises their level of vulnerability to hackers. The possibility of the destruction of human rights due to digital security problems is the biggest human rights challenge of our day. The danger isn't just that some prankster hacks into your smart washer and causes the loop to rise or that your Google Home speaker's microphone is hacked with a warning to subscribe to PewDiePie's YouTube channel. (Yes, it was true.) A hacked smart lock could allow anyone to enter your home unannounced. If you hack only a few smart water heaters, you can trigger widespread blackouts. One insecure computer can cause an entire network to be compromised. As WIRED reported, Fig. 3 "IoT computers have been enlisted into huge botnets to be used for surveillance, bitcoin mining, and attacks on power grids" [5, 13].

The vulnerability to connected devices isn't only because they are connected to the internet, but that the devices manufactures haven't always built their goods with protection as their primary purpose. In 2016, there was a widespread IoT system hack that was caused by a major distributed denial of service (DDoS) attack. The following year, this malware caused virtually every internet-connected computer to become vulnerable to a security flaw. The attack was catastrophic and impossible to mitigate, in part due to the multiplicity of operating systems in the IoT. When a device or phone experiences problems due to a virus or malware, tech developers

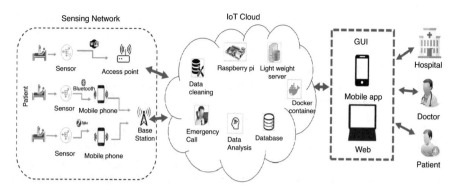

Fig. 3 Types of IoT healthcare networks

are quick to issue a fix. But certain devices like routers and internet-connected door-bells aren't updated to guard against bugs, and many of them weren't made for the same kinds of security protocols as computers. The security researcher made predictions that we'd already discover compromised computers 20 years later [14].

The Internet of Things has also been adapted in the field of electricity and factory automation. According to a Cisco conference call with journalists, "as more ties are made, the importance of companies and the global economy will continue to soar." The eWeek article focuses on a Cisco vision that extends beyond being connected to the Internet and additional connected devices. The Internet is seen as a web of links that contain not only computers and users, but also data and systems, "all that is linked to or crosses over the Internet." Cisco estimates the Internet of things will be worth $14.4 trillion by 2020 [15].

The potential security threats on internet-connected computers continue to be of concern for both firms and users. In 2014, hackers exploited the target system to drain 40 million of their credit cards' funds. How did they find the place? A malicious email was sent to a contractor hired by Target's HVAC system, which had connections to the company's network. The vendor clicked the email and allowed hackers to have remote access, too. In 2019, Amazon was the target of a $5 million class-action lawsuit from users of their internet-connected ring doorbells. Customers told tales of thieves who used their open doorbells to get into homes and commit robbery. The corporation denied responsibility but said that the consumer was to blame for using bad passwords for their account [16].

This security violation motivated California to pass laws that lift the security requirements applied to Internet of Things product manufacturers. The legislation requires some security-related tests in the construction of any computer capable of connecting to the internet. The newly required security measures are a vital first step toward governing the security of internet-connected devices. In 2018, Oregon and California have passed IoT protection legislation, with both laws requiring vendors to make their IoT products as secure as possible.

There are questions about privacy. If they are placed around your house, they are spying on you. Anything on the internet is valuable because much of the information has commercial value. A new analysis of IoT products revealed that 72 of 81 products were sharing data with a third party that was not associated with the product. In addition to your personal belongings, the little bits of your life can be packed and sold to another user. Google and Apple admitted, in 2019, that their smart speakers record users' conversations and that their records are checked by contractors, particularly uncomfortable and personal samples of audio [17].

The next few decades will see the development of the Internet of Everything. Things can feel and react to us differently each time, so even a smart thermostat automatically changes depending on your body temperature or the house locks itself when you get into bed. Your clothes could be linked to embedded sensors that provide a physical reaction to your movements in real time. That is beginning to happen now: In 2017, Google unveiled Project Jacquard, a technology that will enable you to interact with your clothing.

In 2018, there were approximately 23 billion connected devices worldwide. By 2035, the world will have more than 80 billion. Majority of the explosion came from people getting more familiar with an always-on, data-collecting computer that resides in their living room but also from corporations creating novel ways to communicate to the internet. This vision is about more than your home or clothing. You'll have smart workplaces, smart homes, and smart cities too. Smart hospitals will have cameras that will help track hand hygiene and air quality, and towns will have early warning systems for mudslides and other natural disasters. Autonomous cars will link to the internet and become smartened with sensors, and households will monitor their energy use via the Internet of Things. The development of the internet of things could result in new kinds of cyber warfare, allowing a nefarious agent to disable thermostats, pacemakers, and insulin pumps. It could divide our society into those with robot maids, and those without. Or, as Ray Bradbury defined in one short story from 1950, the smart home will be self-sufficient and continue to function without humans [18].

If we want to get where we are going, faster internet connections will be required. Enter "5G." Crazy-fast broadband speeds have long been overpromised and undelivered, but these days, you can see true 5G if you squint. By 2020, the seriousness of the Ebola pandemic sent work and life into cyberspace, and the FCC expanded a plan for upgrading internet connectivity. It may have positive effects on remote work and education but also lead to the wide use of other internet-enabled devices. The Chinese have started testing a 5G version of robotics in hospitals to avoid the spread of infectious diseases such as a novel coronavirus. Also, the internet of things provided funding for health care services this year. Researchers used global positioning services in cell phones to map the progress of the outbreak, public health officials used sensors to detect patients under quarantine, and doctors used internet-connected equipment, including drones and robotics, to transport medicines and check on patients without risking contact.

We need to ensure that these instruments do not interfere with the radio waves. We will need to protect the data that is sent over the frequencies. The Swiss company, Teserakt, is developing a cryptographic implant for IoT devices that can be used to encrypt the data that streams from these devices. And DARPA has also been focusing on improving the protection of their numerous army IoT devices. The mission of that project, delightfully called CHARIOT (Cryptography for Hyper-scale Architectures in a Robust Internet Of Things), is to allow low-cost cryptanalysis to make internet-enabled devices harder to crack. Darpa's technologies are not only for military use but also in GPS, autonomous vehicles, and the Internet. So if the federal government can break IoT encryption, they will probably be able to hack the security on your Home Pod, too.

1.2   Smart Home Dares to Make Sense

The "smart home" has not yet been a viable option. Sure, you might use an app to dim your lights; you might even make a video of your refrigerator. Despite several years of technological progress, the wired home has yet to shed its infancy. It's too costly, too futuristic, too swamped with interoperability and liability issues. Do you need some help with that? Ikea Indeed, it has already begun.

Install the battery using the provided controller, and you are ready to go and play. Smart, but efficient. You can purchase the hub and import apps that can be controlled from your smartphone so that you can monitor them. However, at least you have the option of programming your smart bulbs to be less bright.

"Light fixtures have existed for a long time," said an observer, Bradley Russell of Parks Associates. "It is an appropriate job for them. Most of the lighting applications are architecture, making the brand very design-conscious. Light is a low-cost means to get a smart house."

There is a certain advantage of beginning small and being patient. Controlling the smart home market will help Ikea win the confidence of its consumers. It can be messy, but it's also simple to replace a light bulb. Provided Ikea's dedication to dealing with whatever technology their customers already have installed in their homes [19].

1.3   Improving IoT in Healthcare

Patients admitted to a hospital will benefit from continuous surveillance by IoT-enabled, noninvasive monitoring. This type of system involves sensors that capture detailed physiological data and are then analyzed and stored in the cloud for use by a remote caregiver. It eliminates the practice of making a health provider come by every few days but provides constant automatic access to information (Fig. 4). In this way, it simultaneously increases the standard of treatment by continuous attention and reduces the cost of care by eliminating the requirement for a caregiver to constantly participate in data gathering and research [20].

There are many people everywhere in the globe who will like medical assistance but haven't access to medical services. However, lightweight, efficient wireless

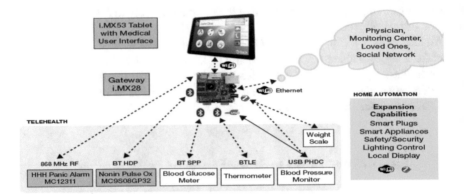

Fig. 4 IoT network topology on remote healthcare monitoring

solutions linked through the Internet of Things have made it possible for surveillance to come to these patients, instead of vice versa. These systems allow for the safe collection of patient health data from a range of devices and then sharing it with medical practitioners who can make suitable decisions for treatment [21].

Early intervention/avoidance involves tracking involved, healthy individuals and using this information to track their well-being (Fig. 5). A senior living alone may want a gadget that can sense a fall or other disruption of their daily activities and notify emergency services or family members. Indeed, an active athlete such as a hiker and biker may profit from such a solution at any age, especially if it is available as a piece of wearable technology [13, 22, 23].

1.4 Why Would the Industry Benefit from Using IoT?

Competitive differentiation – With billions of connected products that collectively share useful knowledge, companies will benefit from industry analysis by benchmarking studies. This allows for a holistic view of various silos within an organization, which advances IoT.

Among the biggest advantages for the industry is innovating for better decision management and data mining. Data created by connected devices can provide companies and management with the valuable decision and innovation information.

So this helps better decision-making. With more contact between stakeholders, higher stock value.

Risk Management – Use of wearable IoT devices is creating greater opportunities to manage employees' fitness, protection, venue, enforcement, and safety requirements. The data can be used to make smarter policies that affect growth, reliability, and organizational performance.

Automation – IoT devices may have several pathways to a finished product. Automation will remove the need for humans to intervene in tasks; therefore, IoT

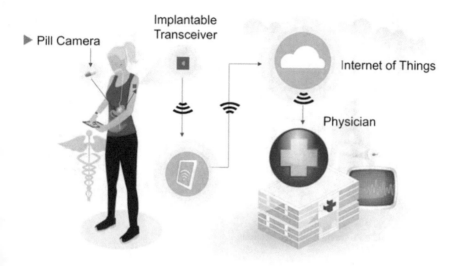

Fig. 5 Framework of IoT based for healthcare receiver for physician

channels can hunt down information offering alternatives to the decision-maker. With this intervention, the patient is more efficient and automated most of the day-to-day tasks which affect progress [16, 24].

1.5 Applied Internet of Things in Healthcare

The Internet of things (IoT) has become synonymous with technologies being embedded in both the consumer and healthcare industries. Organizations of both technology and healthcare that specialize depend upon IoT. With Wi-Fi or Bluetooth powered X-ray machines and wearables including bio-sensors, health practitioners are provided with vital data that aids them in making progress in their profession (Fig. 6).

Despite the incomplete exploration of IoT in the healthcare sector, the speed is high enough to be considered the "Internet of Medical Stuff." Cisco Systems projects that by 2020, there will be 50 billion IoT computers connecting to the internet. This study provides a new perspective on IoT capabilities with detailed information on a variety of variables. In the present day, there are some real advantages of using IoT applications in healthcare systems. The Internet of Things can change healthcare in many ways. Smart devices are wired to the internet, which allows them to view information from everywhere. Hospitals enable IoT technology to improve patient care and safety. Healthcare IoT programs use some other systems [25–27].

IoT technologies enable more efficient processes in healthcare and generate new knowledge for patients (Fig. 7).

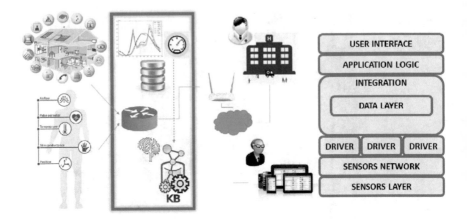

Fig. 6 IoT healthcare services and applications

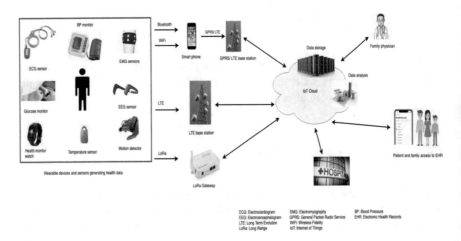

Fig. 7 Devices used in healthcare based on IoT

2 The Advantages of IoT in Healthcare

2.1 Cancer Therapy

Cancer therapy A study conducted in 2018 found that a new medication for head and neck cancer was highly effective in some patients. In the experiment, blood pressure and weight were monitored using a blood pressure cuff and Bluetooth-powered scales along with a nutrition monitoring app. Via a smart-tracking device such as CYCORE, doctors were now able to predict cancer signs, something that was never possible before. The American Society of Clinical Oncology has recognized smart technology's ability to aid in reducing patient treatment burdens and simplifying care delivery [28].

2.2   Diabetes Management

The Internet of Things has created smart devices that continuously track the glucose levels of diabetics (Fig. 8). These are also known as glucose tests administered to treat diabetic patients. Many smart CGMs, like Freestyle Libre and Eversense, transmit knowledge about blood glucose levels to smart tablets, smartphone, and the Apple Watch. Other devices like insulin pens, pens with built-in insulin sensors, and pen caps like InPen, Gocap, and Esysta have the capabilities to automatically record the number, duration, and type of insulin injected in a particular dose [29].

Designing a smart contact lens that could monitor tear glucose would benefit many patients. Treatment for Asthma. Smart technical devices offer insights into the cause of asthma symptoms as well as allowing such medication therapy. Propeller Health is a company that designs and develops smart inhalers. A monitor is used to measure a person's breathing. This is a part of the E-bike app which supports those who are suffering from Chronic Obstructive Pulmonary Disease (COPD) and asthma. Mental Healthcare IoT. In 2003, the World Health Organization observed that over half of the medications prescribed to mental patients were not adminis- tered as directed by the doctor. Proteus also developed pills that will split in the stomach and emit signals which will be picked up by the sensor attached to the body. An extensive amount of trials have shown the effectiveness of drugs to cure various mental disorders [30–32].

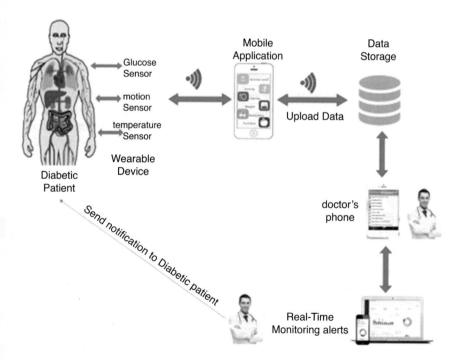

Fig. 8 Diabetes management on collecting and monitoring data

USB-enabled medical devices can bind to the catalogue of diseases for diagnosing purposes. The computer will communicate with the users via the touchscreen to enter data for processing and analysis. When any user enters the data (about different diseases) into the device, it instantly compares and fits the symptoms present on the file, which again matches with the input. This is one of the ways our system reacts to inputs and produces a recommendation for a health condition.

ECG Sensor
The electrocardiogram is a medical instrument that is regularly used to determine the electrical and muscular state of the heart. It can diagnose a variety of heart disorders such as ectopic beats and dramatic tachycardias [33].

Glucometer
When a skin prick is performed, a tiny blood drop is collected from the skin as the meter reads and measures the amount of glucose in the blood, through the glucometer [34].

Airflow Sensor
The nasal airflow sensor is a measure of the volume of air flowing through the nose of a patient needing respiratory support [35].

Blood Pressure Monitor
The blood pressure monitor tracks the systolic and diastolic pressures as well as other blood pressure-related numbers such as pulse rate [36].

3 Challenges for IoT in Healthcare

The benefits of IoT-enabled devices are extensive, but some areas need to be investigated. The Internet of Medical Things has many benefits, but it has some notable drawbacks. These problems require that technological solutions be developed that aren't always available. The key concern concerning the healthcare industry which is making use of IoT technology is security, accessibility, risk, and legislation [37].

4 Conclusion

IoT-based healthcare innovation is already underway, as the explanations in this paper demonstrate. And those are only the tip of the proverbial iceberg, as new use cases begin to rise to meet the urgent need for affordable, efficient treatment. Meanwhile, we are seeing the basic building blocks of automation and machine-to-machine connectivity begin to build out. The service layer completes the infrastructure. IoT technologies are proud to be a part of this movement by offering end-to-end applications for IoT-driven solutions for healthcare, moving on developing guidelines for these solutions and driving IoT-based healthcare creativity for companies willing to understand the advantages of the IoT in healthcare. The importance of IoT in changing the market cannot be understated, nor can it be overstated. Technology and its development can be seen as an inevitable outcome of invention and advances in research. This has brought civilization into conflict at any point of growth. Industry 4.0 and its influence on IoT powered applications would generate a new avenue and effect on a global scale.

References

1. S. Ullah, K.S. Kwak, An ultra-low-power and traffic-adaptive medium access control protocol for wireless body area network. J. Med. Syst. **36**(3), 1021–1030 (2012)
2. A.M. Rahmani, N.K. Thanigaivelan, T.N. Gia, J. Granados, B. Negash, P. Liljeberg, H. Tenhunen, Smart e-health gateway: Bringing intelligence to Internet-of-things based ubiquitous healthcare systems, in *Consumer Communications and Networking Conference, 12th Annual IEEE, Las Vegas,* (2015), pp. 826–834
3. H. Wang, H.S. Choi, N. Agoulmine, M.J. Deen, J.W. Hong, Information-based sensor tasking wireless body area networks in U-health systems, in *Network and Service Management, 2010 International Conference on IEEE, Niagara Falls,* (2010), pp. 517–522
4. W.Y. Chung, Y.D. Lee, S.J. Jung, A wireless sensor network-compatible wearable healthcare monitoring system using integrated ECG, accelerometer and SpO(2), in *Engineering in Medicine and Biology Society, EMBS 2008. 30th Annual International Conference of the IEEE Vancouver,* (2008), pp. 1529–1532
5. M.R. Yuce, Implementation of wireless body area networks for healthcare systems. Sensors Actuators A Phys. **162**(1), 116–129 (2010)
6. J. Ko, C. Lu, M.B. Srivastava, J.A. Stankovic, A. Terzis, M. Welsh, Wireless sensor networks for healthcare. Proc. IEEE **98**(11), 1947–1960 (2010)
7. Y. Chen, W. Shen, H. Huo, Y. Xu, A smart gateway for health care system using wireless sensor network, in *Sensor Technologies and Applications, 2010 Fourth International Conference on IEEE,* (2010), pp. 545–550
8. H. Viswanathan, E.K. Lee, D. Pompili, Mobile grid computing for data- and patient-centric ubiquitous healthcare, in *Proceedings of the 1st IEEE Workshop Enabling Technologies for Smartphone Internet Things,* (2012), pp. 36–41
9. D. Miorandi, S. Sicari, F. De Pellegrini, I. Chlamtac, Internet of things: Vision, applications and research challenges. Ad Hoc Netw. **10**(7), 1497–1516 (2012)
10. X.M. Zhang, N. Zhang, An open, secure and flexible platform based on Internet of things and cloud computing for ambient aiding living and telemedicine, in *Computer and Management, 2011 International Conference on IEEE, Wuhan,* (2011), pp. 1–4
11. W. Wang, J. Li, L. Wang, W. Zhao, The Internet of Things for resident health information service platform research, in *Proceedings of IET International Conference on Communication Technology and Application,* (2011), pp. 631–635
12. L. Yang, Y. Ge, W. Li, W. Rao, W. Shen, A home mobile healthcare system for wheelchair users, in *Proceedings of the IEEE 18th International Conference on Computer Supported Cooperative Work Design, Hsinchu, Taiwan,* (2014), pp. 609–614
13. A. Alaiad, L. Zhou, Patients adoption of WSN-based smart home healthcare systems: An integrated model of facilitators and barriers. IEEE Trans. Prof. Commun. **60**(1), 4–23 (2017)

14. V.M. Rohokale, N.R. Prasad, R. Prasad, A cooperative Internet of Things (IoT) for rural health-care monitoring and control, in *Wireless Communication, Vehicular Technology, Information Theory and Aerospace & Electronics Systems Technology, 2011 2nd International Conference on IEEE*, (2011), pp. 1–6
15. D.T. Matt, G. Arcidiacono, E. Rauch, *Applying Lean to Healthcare Delivery Processes – a Case-based Research* (Int. J. Adv. Sci. Eng. Inf. Technol, 2018)
16. M.G.N. Musoke, Health information in Uganda, in *Informed and Healthy*, (2016)
17. S. Sitepu, H. Mawengkang, Irvan, Modeling an integrated hospital management planning problem using integer optimization approach. J. Phys. Conf. Ser. (2017)
18. M.Y. Lin, K.S. Chin, K.L. Tsui, A weighted multi-objective therapist assignment problem in Hong Kong, in *IEEE International Conference on Industrial Engineering and Engineering Management*, (2016)
19. A.M.H. Pardede et al., A framework for sharing patient data and service facilities using Ipv6 technology to support the advancement of smart health care technologies A. J. Phys. Conf. Ser. (2019)
20. S. Sitepu, H. Mawengkang, A two-stage stochastic optimization model of hospital nursing staff management problem. Int. J. Adv. Res. Comput. Eng. Technol. **4**(1), 44–47 (2015)
21. P.M. Cunningham et al., Implications of baseline study findings from rural and deep rural clinics in Ethiopia, Kenya, Malawi and South Africa for the co-design of mHealth4Afrika, in *GHTC 2016 – IEEE Global Humanitarian Technology Conference: Technology for the Benefit of Humanity, Conference Proceedings*, (2016)
22. Z. Wang, Z. Yang, T. Dong, A review of wearable technologies for elderly care that can accurately track indoor position, recognize physical activities and monitor vital signs in real time. Sensors (Switz.) (2017)
23. S.P. Mohanty, U. Choppali, E. Kougianos, Everything you wanted to know about smart cities. IEEE Consum. Electron. Mag. **5**(3), 60–70 (2016)
24. S. Purkayastha, J.W. Gichoya, A.S. Addepally, Implementation of a single sign-on system between practice, research and learning systems. Appl. Clin. Inform. (2017)
25. V. Chichernea, The use of decision support systems (dss) in smart city planning and management. J. Inf. Syst. Oper. Manag., 1–14 (2014)
26. B.B. Nasution et al., Forecasting natural disasters of tornados using mHGN, in *IFIP Advances in Information and Communication Technology*, (2017)
27. A.M.H. Pardede et al., Framework for patient service queue system for decision support system on smart health care. Int. J. Eng. Technol. **7**(2.13), 337–340 (2018)
28. I. Tulus, Z. Sefnides, Sawaluddin, Suriati, M. Dwiastuti, Modeling of sedimentation process in water, in *2nd International Conference on Computing and Applied Informatics 2017*, (2018), pp. 1–5
29. M.K.M. Nasution, Modelling and simulation of search engine, in *International Conference on Computing and Applied Informatics 2016*, (2017), pp. 1–8
30. A. Alsalemi et al., Developing cost-effective simulators for patient management: A modular approach, in *2017 Fourth International Conference on Advances in Biomedical Engineering (ICABME)*, (2017), pp. 1–4
31. K. Dorling, J. Heinrichs, G.G. Messier, S. Magierowski, Vehicle routing problems for drone delivery. IEEE Trans. Syst. Man, Cybern. Syst. (2017)
32. L. Gu, D. Zeng, S. Guo, A. Barnawi, Y. Xiang, Cost efficient resource management in fog computing supported medical cyberphysical system. IEEE Trans. Emerg. Top. Comput. (2017)
33. A.M.H. Pardede et al., Smart health model with a linear integer programming approach. J. Phys. Conf. Ser. (2019)
34. D. Niyigena, C. Habineza, T.S. Ustun, Computer-based smart energy management system for rural health centers, in *Proceedings of 2015 IEEE International Renewable and Sustainable Energy Conference, IRSEC 2015*, p. 2016

35. S.S. Savanth, K.N.R.M. Babu, Hospital queuing-recommendation system based on patient treatment time, in *Proceedings of the 2017 International Conference on Intelligent Computing and Control Systems, ICICCS 2017*, p. 2018
36. J. Chen, K. Li, Z. Tang, K. Bilal, K. Li, A parallel patient treatment time prediction algorithm and its applications in hospital queuing-recommendation in a big data environment. IEEE Access (2016)
37. X. Chen, L. Wang, J. Ding, N. Thomas, Patient flow scheduling and capacity planning in a smart hospital environment. IEEE Access **4**, 135–148 (2016)

Automated and Curated Sack Count Leveraging Video Analysis on Moving Objects

Ritin Behl, Harsh Khatter, Prabhat Singh, Garima Bhardwaj, and Prateek Chaturvedi

1 Introduction

From the advances in car technology to the various features that help drivers in modern cars, the long way of automotive technology is becoming increasingly sophisticated with the Laptop vision that results in greater and easier. At the same time, infrastructure transportation providers and resources will increase their reliance on CV to improve their safety and efficiency in transportation. The concept of computer in this way helps to solve critical problems at both the transit level, the consumer-level, and the infrastructure providers [1].

The growing demand for limited transportation infrastructure is constantly growing and leading to traffic delays, accidents, and traffic congestion have also had serious economic consequences. Technological advances such as the concept of computers play a key role in solving these types of problems in efficient and effective ways such as traffic charging, incident detection and management, traffic

R. Behl
Department of Information Technology, ABES Engineering College, Ghaziabad, India

ABES Engineering College, Ghaziabad, India
e-mail: ritin.behl@abes.ac.in

H. Khatter · P. Singh
ABES Engineering College, Ghaziabad, India

Department of Computer Science and Engineering, ABES Engineering College, Ghaziabad, India
e-mail: harsh.khatter@abes.ac.in; prabhat.singh@abes.ac.in

G. Bhardwaj · P. Chaturvedi (✉)
AUGN, ABES Engineering College, Ghaziabad, India

Amity University Greater Noida Campus, Greater Noida, Uttar Pradesh, India
e-mail: gbhardwaj@gn.amity.edu; pchaturvedi1@gn.amity.edu

© The Author(s), under exclusive license to Springer Nature Switzerland AG 2021
S. Awasthi et al. (eds.), *Artificial Intelligence for a Sustainable Industry 4.0*,
https://doi.org/10.1007/978-3-030-77070-9_17

monitoring, traffic monitoring and vehicle control and many more alternatives; and a wide range of applications that computer vision technology can support. Self-help assistance programs are being rolled out at increasing prices, these programs will begin to shift their role from one aid to another to facilitate decisions as they are related to safety but also the power that goes up to connect the infrastructure [2].

The aim of the independent acquisition programs is to reduce traffic congestion, to produce less expensive cars and emergency equipment, to improve public safety, to reduce environmental impacts, to improve mobility data, and so on. The technology helps regions, cities, and towns at the national level to meet the growing demands of a better travel system. The effectiveness of an independent system is largely based on the efficiency and completeness of the acquisition technology [3, 4].

Car detection by video cameras is one of the most attractive and inaccessible technologies that require large data collection and use of vehicle control and management systems. Object discovery is also the basis for tracking. Proper discovery of an object results in better tracking of the object. Modern computer-controlled systems have more complex access requirements than those used by conventional traffic controllers as one of the road signs, for which many explosive devices are made [5].

2 Problem Statement

Often it happens, human observation is required to keep track of objects unloaded from the truck to the warehouse or loaded from the warehouse into the trucks. There is a possibility of errors. To solve this problem a web application will be designed which will count the number of sacks that are loaded into the truck or unloaded from the truck. A camera would be used which will be monitoring the sacks that are loaded and unloaded, and it would be easier to find if the sacks loaded are equal to the number of sacks unloaded or not. With the advancement in video analytics, it is feasible to remove the human observation and replace it with a camera-based video analytics solution by object detection method [6–8].

This sack counting system is one of the smart solutions for the transportation industry. There is a chance of error and false results in human observation. The goal of the sack counting system backend engine is to process the image frames passed to it via the image processing algorithm according to architectural diagram. The backend engine is trained as such, that, it processes the image frames which was found in the frame folder and predicts the sack count using predictive algorithm and returns live counter of the sacks loaded on to the truck. The engine used based on Tensorflow v2 frame work, and the image frames are fetched from the live video recording from the CCTV camera recording the video [9, 10].

The development of the project will flow in steps one after another such as selecting a model, adapting an existing dataset that creates and annotate own dataset, modifying the model configure the file, trains the model and saving the model, moves the model in another piece of software, and finally displays the result of backend engine on the custom-developed User Interface in the form of web application.

The data would be collected from live camera feed, and then the collected video would be converted into frames. The frames will be processed then, to extract the required information, i.e., the sacks. The model building algorithm would learn from the frames about the color, shape, size, and position of the sacks. Then the model will be tested against some data which is not used for training.

To count the number of sacks a region of interest (ROI) will be determined. The sacks crossing beyond the ROI would be counted and the counter will be incremented by 1. Initially, the counter would be set to zero. The counter will be refreshed every time the model is run again after terminating the model once. The result output is the command prompt screen of the sack count system containing the number of sacks either loaded or unloaded from the truck. Also the annotated frames with confidence score is also displayed on the output window [11].

2.1 Related Work

Object acquisition is a common term used for CV techniques that classify and locate objects in a photograph. The discovery of the modern object depends on the use of the CNN network. There are many programs suitable for Faster R-CNN, R-FCN, Multibox single Shot Detector (SSD), and YOLO (You Just Look Once). The original R-CNN methods work with neural net classifier performance on plant sample images using external suggestions (samples are planted with external box suggestions, with the extraction feature performed on all planted samples). This method is very expensive due to the many crops, the fast R-CNN which is able to reduce the scale by doing the feature once only in the whole image and uses the crop on the lower parts of the sample [4, 5].

Faster R-CNN goes a step further using extruded materials used to create agnostic box suggestions, i.e., a single-release feature for the whole image, no external box suggestions. R-FCN is similar to Faster R-CNN, but feature harvesting is done on different types of enhancement efficiency. The advantage of the simplicity is that the YOLO model is faster compared to the Faster R-CNN and SSD which reads representation. This increases the error rate of the area and YOLO does not work well with images with a new feature rating but reduces the false rating. This will speed up real-time applications. This comes with the price of reduced accuracy [6–8].

SSD with MobileNet refers to the model in which the SSD and the type of SoftNet installer. Accurate tradeoff speed and many other modern systems for finding something that require real speed. Methods like the YOLO and SSD work faster, this often comes with a decrease in predictability, while Faster R-CNN models achieve higher accuracy but are more expensive to run. The cost per model speed depends on the application. SSD works compared to more expensive models like Faster R-CNN; in smaller images its performance is much lower [11].

3   Proposed Approach

The objective is to construct detection and counting model such that it will increase the efficiency of the loading/unloading process in trucks and also increment the efficiency of counting over human observation and lessen the chances of human forgery in the transportation industry. But before constructing the model, the need to study the present scenario and understand the factors or parameters that affect the statistics. The objective was pursued as given in the diagram.

To achieve the objective, the focus was put on data collection and analysis and model research. Also the detection model or implementation part is also a goal. The parameters and their prioritized values are decided only after proper analysis. For this, a dataset as large as possible is required. The discovery of the TensorFlow framework is used to build an in-depth learning network that helps solve object acquisition problems. This includes a collection of pre-defined retrained models trained in the COCO database, as well as the Open Pictures Dataset. These types can be used humbly if there is interest in categories on this data only. They are also useful for implementing models when trained in novel data.

3.1   Model Development

Since the proposed model is based on custom object detection, there will be basic 6 steps to achieve the goal.

3.1.1   Step 1: Problem Definition

This is the first step in finding an item. Defining a problem carefully requires an understanding of how things will be used, who needs the items, and how the adoption function fits into an organization that needs to predict. Job description, outsourcing, decision-making of a project, general discussion of trade-offs (accuracy vs speed), and setting a project codebase are the basic and fundamental requirements of a particular problem.

3.1.2   Step 2: Gathering Information

There are at least two types of information required: (a) image data and (b) live camera feeds. Usually, it will be difficult to get enough live video data to be able to fit a good image capture model due to the large background noise. However, sometimes, older data will not be as useful due to changes in the upload method. For example, today, conveyor belts are more popular than handicrafts.

3.1.3   Step 3: Training Data

Always start by labelling details. Training data are images collected as samples and interpreted to train deep neural networks. With the discovery of something, there are a lot of methods used to fix and quote data for training purposes. Many popular ways to distribute databases such as Pascal containing labelling tools.

3.1.4   Step 4: Feature Extraction

Feature extraction is a basic step used by automated methods based on machine learning methods. Its main purpose is to extract useful features or important components from the data, which the computer can detect to calculate values in input images. An element is defined as the work of 1 or more dimensions that describe a particular asset (such as color, texture, or texture) of the whole image or layer or object.

3.1.5   Step 5: Classification

The feature release feature provides a collection of records seen by a group of features x and a class label y. Aimed at describing a class model that includes a recording class label. Image classification follows steps such as pre-processing, component fragmentation, feature extraction, and segmentation.

3.1.6   Step 6: Testing Data

There is test data within the file which holds to 30% of the full dataset. This test data isn't being employed for training the model. Even after training the model, it is required to check the model on some dataset. It's very significant to understand the accuracy of the full model and confidence score of a selected test run on a trained model.

4   Proposed Methodology

The goal of the sack counting system is to process the image frames passed to it via the image processing algorithm mentioned in the above diagram. The engine is trained as such, that, it is processes the image frames found in the frame folder and predicts the sack using a predictive algorithm and returns the live count of the sacks loaded on to the truck. The engine used the OpenCV and the image frames are fetched from the live recording from the CCTV camera recording the video. The external interface of this engine begins with the collection of frames saved in a .csv folder (Fig. 1).

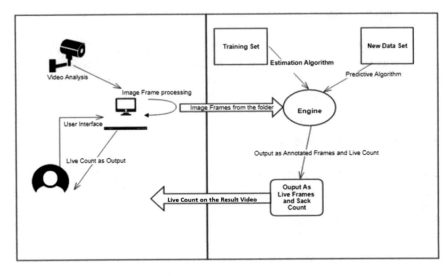

Fig. 1 Architecture diagram of sack count system

The frames are extracted through a live video stream of Camera installed at the work area. Then the OpenCV engine takes the frames one by one from the folder and starts analyzing through a command-line interface. The engine then identifies the correct frames and increments the counter, now the frames become annotated. Finally, the counter and annotated frames, the desired result will be the output video. The annotated result video will become the output. The internal interface will be decomposed into two parts: Train and Test.

The train part will be used to train the model for detecting the frame with a sack, initially making of the dataset is done by own and divided randomly into training and validation sets. It contains images frames along with the label of sack marked on that frame using labelling annotation. Here, the engine will call the main function which in turn calls the train function which first initializes the parameters randomly and updates weights at each iteration up to some minimum cost. The model hyperparameters according to the variance and bias provided by the model are tuned. It is a continuous process to make a model robust to new images and provide more accuracy. This process will train the model, and final parameters can be used to predict the new frame.

In the test part or operational phase, the main function will be called, which refers to predict function. The predict function will check the image frame according to the trained model parameters. If the sack is present in an image frame, then the result would be there, in the form of an output result video which will show the number of sacks loaded/unloaded in truck and time and date of the loading/unloading in truck.

4.1 System Design

1. Data Flow Design

The data flow diagram (DFD) shows the flow of information for any process or system. Data flow diagrams represent systems and processes that will be difficult to define in textual format (Fig. 2).

4.2 Sequence Diagram

A sequence diagram shows the object interactions that are arranged in time sequence. It would describe the classes and objects which are involved in the sequence of messages and exchange between the objects needed to carry out the functionality of the structure (Fig. 3).

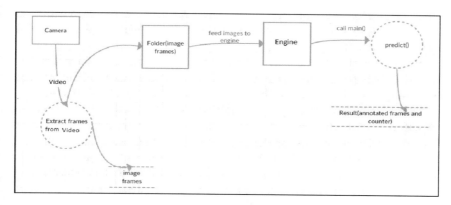

Fig. 2 Data flow diagram of the proposed method

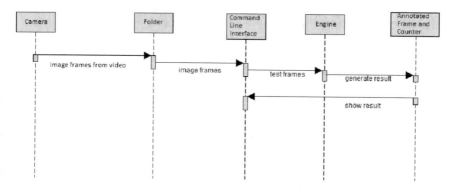

Fig. 3 Sequence diagram of the proposed approach

5 Implementation Details

5.1 TensorFlow framework

TensorFlow is an open-source framework used to integrate and build great machine learning programs. It is created by the Google Brain team. It incorporates mechanical and in-depth learning methods as well as neural network models and algorithms and makes them useful in the form of a standard platform. It uses Python as a programming language that provides a simpler and better complete API for building applications using a framework and also aids in the development of those applications in greater C ++ functionality. TensorFlow helps with prediction of production at a certain level, with the same models used to train a deep neural network. It can train and operate deep neural networks. For example, neural networks used for image recognition, handwriting digitization, duplicate neural networks, embedding words, sequential machine translation models, etc.

Python language is supported by TensorFlow to provide all framework functions. Python is a simple structured language. It is easy to read and work on and offers useful options for explaining how high-end manufacturers can be integrated. Nodes and tensors are Python objects, and the provided Python applications are TensorFlow applications themselves.

In Python, real mathematical functions are not performed. C ++ phones that are known for high-performance work with conversion libraries available for language processing, as well as PDE (which is part of the separate classification) used.

TensorFlow allows developers to create dataflow graphs. Dataflow graphs are defined as structures that show how data moves across a graph, or processing nodes in a series. Mathematical performance is represented by nodes in the graph. Each edge between the areas is a multi-distance data system called tensor.le in the TensorFlow framework. Advanced scams are provided by Python which simply directs traffic between pieces and connects loopholes at intersections.

TensorFlow applications can be distributed to any target and user-friendly, be it machine, cloud collection, iOS and Android devices, or CPUs or GPUs. If you use Google's own cloud, you'll be able to use TensorFlow in the Google Silicon TensorFlow Processing Unit (TPU) to keep up the speed.

Output models of the TensorFlow framework, however, will be used on any device that can be used to provide prediction. TensorFlow 2.0 analyzes the framework in some ways that support user feedback, to make it easier to find that (e.g., by using the simple Keras API for training) and to do more. Distributed training is very complex to run due to the replacement API, and support for TensorFlow Lite enables us to deploy models in a large platform style. However, the code written for previous versions of TensorFlow must be rewritten only occasionally, and sometimes significantly so that the maximum benefit of the latest TensorFlow 2.0 features can be established.

One of the great advantages of TensorFlow is that it offers the advantages of machine learning development. Instead of managing the nitty-gritty details of using algorithms, or determining the appropriate ways to capture the first result of a task in another installation, the developer can work specifically on the general concept of the application. TensorFlow takes care of the small print behind the scenes.

Also, TensorFlow offers one more benefit for developers who need to fix a problem and take TensorFlow details. In the framework, there is a mode of eager use that allows one to test and modify the performance of each graph especially and without stitching, instead of naming the graph as one opaque object and testing it all at once. And the TensorBoard visualization Suite allows one to view and select graph-driven interactive dashboard.

TensorFlow also received a chance for Google support. Google has not only allowed for faster growth after the project but also has created many important TensorFlow methods that make it much easier to use and operate. For example, the abovementioned TPU silicon for faster performance in the Google cloud, your browser and the desired size for the frame, the hub for online frame sharing, and much more.

5.2   Faster R-CNN Algorithm

R-CNN was launched in 2014 and has attracted a lot of interest in the computer vision community. The whole idea behind R-CNN is to implement a search engine selected to propose 2000 Region-and-Interest (ROI), which is then fed into the Convolutional Neural Network (CNN) to collect features. These features have been used to classify photographs and their object boundaries using SVM (support vector machine) and regression methods. For a more detailed explanation, see this section. And this approach was quickly followed by R-CNN, which became a faster and better way to identify an object. Fast R-CNN uses a ROI pooling system that shares features throughout the image and uses a modified type spatial pyramid pooling method to efficiently capture features over time. In the case of fast R-CNN, it is still slow because it has SS to handle, which is computationally very slow. The test time to test Fast R-CNN takes from 47 s to 0.32 s and 2 s to generate 2000 ROIs. It adds up to 2.3 s per image (Fig. 4).

The shortcomings of the above two algorithms paved the way for researchers to quickly come up with R-CNN, where the test time for each image with field resolutions was only 0.2 s. This is due to the latest approach given the completely different model used for end-to-end training.

There are two proposals for fast R-CNN, such as field proposal networks (RPNs) used to generate field proposals and networks that use these found resolutions to identify objects. The big difference here with Fast R-CNN is that the latter fields use selective search to obtain proposals. The time and cost of building field resolutions is much lower in RPNs than in selected searches, as RPNs share very important calculations with the detection network. However, RPNs call field boxes anchors and propose them as objects.

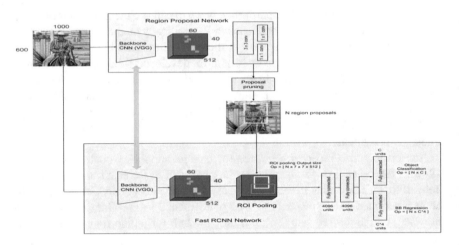

Fig. 4 Faster R-CNN algorithm

The Faster R-CNN Object Detection Network feature is designed with an extraction network that can often use a blocked traditional neural network. After, this two subnetworks can be trained. The primary may be the Region Proposal Network (RPN), which is accustomed to generating object conclusions. The latter is therefore used to estimate the specific class of a given article. The primary variant for fast R-CNN is RPN, which is added after the final composite layer. It is often trained to provide field conclusions without the need for external mechanisms such as selective search. ROI pooling can be used as an upstream classification and bounding box resistor similar to the one used in Fast R-CNN.

5.3 Inception Model

The beginning may be the deeply controversial neural specification introduced in 2014. It won the ImageNet Large-Scale Visual Recognition Challenge (ILSVRC14). It was mostly developed by Google researchers.

In convolutional neural networks (CNNs), the outer part of the work is performed at the appropriate layer for use between the most common options (1×1 filter, 3×3 filter, 5×5 filter, or maximum-pooling). We all want to find the right local structure and replicate it spatially (Fig. 5).

As each of these inception modules is stacked on top, their output correlation statistics vary: as the properties of the upper extraction are captured by the higher layers, their spatial concentration of 3×3 and k reduction is estimated, indicating the ratio. 5×5 resolutions increase as we move to higher layers.

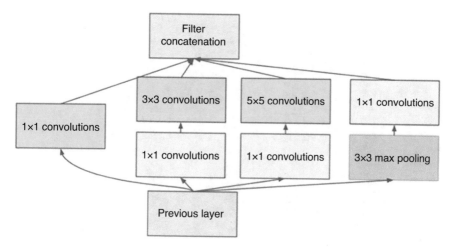

Fig. 5 Inception model

However, the computational cost of such an answer would increase enormously. For this reason, within the figure, diffusion reduction is used as diffusion reduction methods by 1 × 1 convolution.

The installation module aims to act as a "multi-level feature extractor" by calculating the 1 × 1, 3 × 3 and 5 × 5 additives in the homogeneous modules of the network – the output of those filters being then stacked together. Channel size and before feeding in the bottom layer of the network.

The original version of this architecture was called GoogleNet, but later versions were called Inception VN, where n refers to the version number entered by Google. The Inception V3 architecture included in the Kears Core comes from a later publication by Szegedi. Inception Architecture for Computer Vision (2015) proposing an update of the Inception module to further increase image net classification accuracy. Initially the V3 weighed 93 MB, which is smaller than both VGG and ResNet.

5.3.1 Data Collection

The classifier is built in taxonomy to identify sacks. Dataset is a very important thing in creating taxonomy. This may be the basis for our classification, on which object detection ceases. Various and varied images containing objects are collected. Then the directory name image inside the research directory is created. Now, 80% of the photos are stored in the train directory and 20% of the images are stored in the test directory inside the pictures directory. We have collected 174 images in the train directory and 35 images in the test directory (Figs. 6 and 7).

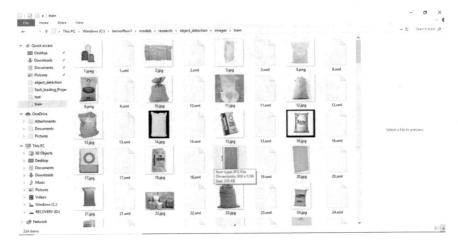

Fig. 6 Train directory

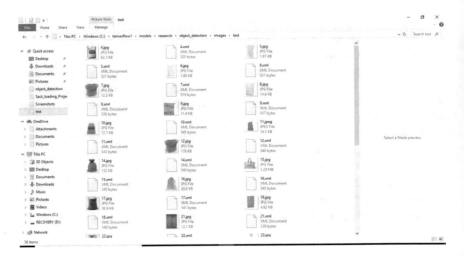

Fig. 7 Test directory

5.3.2 Labelling the Dataset

Open the Labelling tool and start drawing rectangular boxes on the image where the subject is. Label them with the appropriate name as shown in Fig. 8. Save each image created by labelling an xml file with the corresponding image name as shown in Fig. 9.

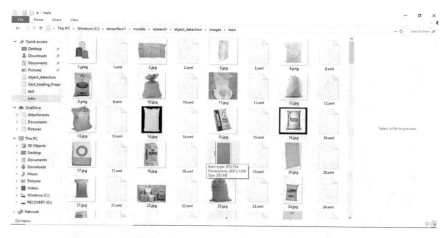

Fig. 8 Train directory with xml file

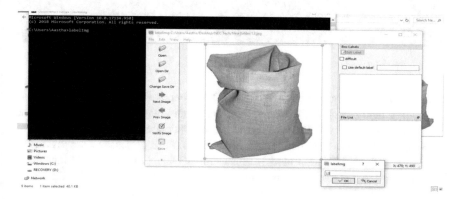

Fig. 9 Labelling tool

5.3.3   Generating TensorFlow Records for Training

For this step, we want to create TFRecords that can be displayed as an input file to train the subject detector. Xml_to_csv.py and Gener_tfrecord.py codes are given. Create test.csv and train.csv files in the xml_cs_csv.py images folder. Then, create files for recording in TensorFlow by executing the appropriate commands from the Object_Detection folder.

5.3.4   Configuring Training

Create a replacement directory called Training in the Object_Detection directory. Use the text editor to create the restore file and place it as a label map in the training directory. The label tells the map instructor what each object is by defining the

```
INFO:tensorflow:global step 48173: loss = 0.0575 (0.945 sec/step)
INFO:tensorflow:global step 48174: loss = 0.0039 (0.948 sec/step)
INFO:tensorflow:global step 48175: loss = 0.0241 (0.948 sec/step)
INFO:tensorflow:global step 48176: loss = 0.0403 (0.945 sec/step)
INFO:tensorflow:global step 48177: loss = 0.0089 (0.946 sec/step)
INFO:tensorflow:global step 48178: loss = 0.0125 (0.947 sec/step)
INFO:tensorflow:global step 48179: loss = 0.0253 (0.942 sec/step)
INFO:tensorflow:global step 48180: loss = 0.0135 (0.952 sec/step)
INFO:tensorflow:global step 48181: loss = 0.0028 (0.945 sec/step)
INFO:tensorflow:global step 48182: loss = 0.0049 (0.945 sec/step)
INFO:tensorflow:global step 48183: loss = 0.0034 (0.950 sec/step)
INFO:tensorflow:global step 48184: loss = 0.0031 (0.949 sec/step)
INFO:tensorflow:global step 48185: loss = 0.0044 (0.949 sec/step)
INFO:tensorflow:global step 48186: loss = 0.0713 (0.948 sec/step)
INFO:tensorflow:global step 48187: loss = 0.0072 (0.947 sec/step)
INFO:tensorflow:global step 48188: loss = 0.0312 (0.952 sec/step)
INFO:tensorflow:global step 48189: loss = 0.0256 (0.950 sec/step)
INFO:tensorflow:global step/sec: 0.979725
INFO:tensorflow:global step 48190: loss = 0.0111 (1.105 sec/step)
INFO:tensorflow:Recording summary at step 48190.
INFO:tensorflow:global step 48191: loss = 0.0084 (1.091 sec/step)
INFO:tensorflow:global step 48192: loss = 0.0080 (0.949 sec/step)
INFO:tensorflow:global step 48193: loss = 0.0410 (0.948 sec/step)
```

Fig. 10 Training model

mapping of the advanced name to the class ID number. Now, add the content to class-map.pbtxt, then switch to the format to create a label map for your taxonomy. The label map ID number should be as defined in the Generate_tfrecord.py file. We need a model algorithm to train our classification. During this project, we are visiting to use the faster_rcn_inception model. TensorFlow's Object Detection API includes a large number of models. Now, open the file using the text editor and make the necessary changes to the fast_rcnn_inception_v2_pets.config file saved in the directory.

5.3.5 Training Model

At the situation object_detection/legacy/find the file train.py. Open the object detection directory and copy the train.py file and paste it within the same directory. Run the subsequent command to begin training the model in the object detection folder itself. It takes around 1–2 min to start the setup before the training begins. The training starts, and it looks like given in Fig. 10.

6 Conclusion

The work done so far is concentrated on the collection image data and labelling from and training the model for object detection in an image. Also the focus was on different algorithms that are there and their comparative study for analysis of video data. The detection model will take these resulting parameters as input and produce a prediction. The goal to convert this analysis into an accurate detection and counting using these object detection and tracking algorithms has been achieved.

The framework used, that is the TensorFlow framework, in this project for the application of object count can be further extended and optimized through various means. It can make simple but significant modifications in the model while training and during data pre-processing like increasing the training size, more epochs in training, and bigger size images for improving the accuracy. Also we can implement this whole idea on mobile devices using MobileNet (included in TensorFlow framework) which provides better accuracy.

References

1. J. Huang, V. Rathod, C. Sun, M. Zhu, A. Korattikara, A. Fathi, I. Fischer, Z. Wojna, Y. Song, S. Guadarrama, Speed/accuracy trade-offs for modern convolutional object detectors. arXiv preprint (2016) arXiv:1611.10012
2. J. Redmon, S. Divvala, R. Girshick, A. Farhadi, You only look once: Unified, real-time object detection, in *Proceedings of the IEEE Conference on Computer Vision and Pattern Recognition*, (2016), pp. 779–788
3. W. Liu, D. Anguelov, D. Erhan, C. Szegedy, S. Reed, C. Fu, A.C. Berg, Ssd: Single shot multibox detector, in *European Conference on Computer Vision*, (Springer, 2016), pp. 21–37
4. B. Galitsky, A content management system for Chatbots, in *Developing Enterprise Chatbots*, vol. 1, (Springer, 2019), pp. 253–326
5. N. Ali, M. Hindi, R.V. Yampolskiy, Evaluation of authorship attribution software on a chatbot corpus, in *International Symposium on Information, Communication and Automation Technologies*, (2011)
6. E. Go, S.S. Sundar, Humanizing chatbots: The effects of visual, identity and conversational cues on humanness perceptions. Comput. Hum. Behav. **97**, 304–316 (2019)
7. P. Singh, H. Khatter, S. Kumar, Evolution of software-defined networking foundations, in *Evolution of Software-Defined Networking Foundations for IoT and 5G Mobile Networks*, vol. 1, (2021), pp. 98–112
8. H. Khatter, B.M. Kalra, A new approach to Blog information searching and curating, in *Proceedings to Sixth International Conference on Software Engineering CONSEG 2012, Indore, India*, (2012), pp. 1–6
9. H. Khatter, A.K. Ahlawat, Analysis of content curation algorithms on personalized web searching, in *Proceedings of the International Conference on Innovative Computing & Communications (ICICC)*, (New Delhi, 2020), pp. 1–4. https://doi.org/10.2139/ssrn.3563374
10. H. Khatter, M.C. Trivedi, B.M. Kalra, An implementation of intelligent searching and curating technique on Blog Web 2.0 tool. Int. J. U E-Serv. Sci. Technol. **8**(6), 45–54 (2015)
11. H. Khatter, A.K. Ahlawat, An intelligent personalized web blog searching technique using fuzzy-based feedback recurrent neural network. Soft Comput. **24**(12), 9321–9333 (2020). https://doi.org/10.1007/s00500-020-04891-y

Index

Printed in the United States
by Baker & Taylor Publisher Services